Natural Product Chemistry at a Glance

T0225976

Natural Product Chemistry at a Glance

Dr Stephen P. Stanforth

School of Applied Sciences, Northumbria University

Blackwell
Publishing

© 2006 Dr Stephen P Stanforth

Editorial offices:
Blackwell Publishing Ltd, 9600 Garsington Road, Oxford OX4 2DQ, UK
    Tel: +44 (0)1865 776868
Blackwell Publishing Inc., 350 Main Street, Malden, MA 02148-5020, USA
    Tel: +1 781 388 8250
Blackwell Publishing Asia Pty Ltd, 550 Swanston Street, Carlton, Victoria 3053, Australia
    Tel: +61 (0)3 8359 1011

The right of the Author to be identified as the Author of this Work has been asserted in accordance with the Copyright, Designs and Patents Act 1988.

All rights reserved. No part of this publication may be reproduced, stored in a retrieval system, or transmitted, in any form or by any means, electronic, mechanical, photocopying, recording or otherwise, except as permitted by the UK Copyright, Designs and Patents Act 1988, without the prior permission of the publisher.

First published 2006 by Blackwell Publishing Ltd

ISBN-10: 1-4051-4562-5
ISBN-13: 978-1-4051-4562-6

Library of Congress Cataloging-in-Publication Data

Stanforth, Stephen P.
    Natural product chemistry at a glance : Stephen P. Stanforth.– 1st ed.
        p.   cm.
    Includes index.
    ISBN-13: 978-1-4051-4562-6 (pbk. : alk. paper)
    ISBN-10: 1-4051-4562-5 (pbk. : alk. paper)
    1. Natural products.   I. Title.

    QD415.S725 2006
    547′.7–dc22   200503168

A catalogue record for this title is available from the British Library

Set in 10/11 pt Times
by TechBooks, New Delhi, India

The publisher's policy is to use permanent paper from mills that operate a sustainable forestry policy, and which has been manufactured from pulp processed using acid-free and elementary chlorine-free practices. Furthermore, the publisher ensures that the text paper and cover board used have met acceptable environmental accreditation standards.

For further information on Blackwell Publishing, visit our website:
www.blackwellpublishing.com

Contents

Contents

Introduction

About this Book

This book has been designed to introduce the reader to the underlying concepts of natural product biosynthesis and in particular to highlight the similarities between many organic and biological reactions. The material covered in this book has therefore been chosen to reinforce the reader's understanding of organic reaction mechanisms and provide a link to their biological counterparts. Each section within the book considers a key biosynthetic building block and discusses its assembly into a variety of natural products. Within most sections there are a number of problems for the reader to attempt.

The material covered in this book is based upon an open-learning module that was developed with funding from 'Project Improve' (now the LTSN[1] in physical sciences) whose support is gratefully acknowledged. The author has used this module for teaching natural product biosynthesis to undergraduate students at Northumbria University since 1998. The original material is still freely available (see further reading) but has been significantly extended and amended to produce this book.

---

[1] LTSN is the learning and teaching support network in physical sciences.

1.1 Primary and Secondary Metabolites

The origin, properties and purpose of natural products has fascinated researchers for many years. Natural products are generally divided into two broad classes, primary metabolites and secondary metabolites. The former class comprises molecules that are essential for life; principally proteins, carbohydrates, fats and nucleic acids, and these molecules are produced by metabolic pathways common to most organisms. Primary metabolic pathways are therefore concerned with processes that synthesize, degrade and interconvert these primary metabolites. In contrast, secondary metabolites are usually of relatively limited occurrence and are often unique to a particular species. Secondary metabolites are produced from a few key intermediates of primary metabolism.

Acetyl coenzyme A **1.1** (Figure 1.1) is the biological equivalent of acetate ($CH_3CO_2{}^-$) and this compound is produced from glucose in several steps. Acetyl coenzyme A is an important intermediate in primary metabolism but it is also a key intermediate in the formation of many classes of secondary metabolites. Fatty acids (e.g. stearic acid **1.2**), compounds derived from fatty acids (e.g. prostaglandin PGF$_{2\alpha}$ **1.3**), many phenolic compounds (e.g. orsellinic acid **1.4** and griseofulvin **1.5**) are all derived from acetyl coenzyme A. Section 2 introduces the underlying chemistry of acetyl coenzyme A and sections 3 and 4 illustrate how acetyl coenzyme A gives fatty acid derivatives and polyketide-derived products respectively from multiple acetyl coenzyme A units. Polyketides are compounds that possess alternating carbonyl and methylene groups i.e. ($-CO-CH_2-$)$_n$ with each unit originating from an acetyl coenzyme A molecule. By using other coenzyme A derivatives in place of acetyl coenzyme A, Nature has increased the diversity of natural

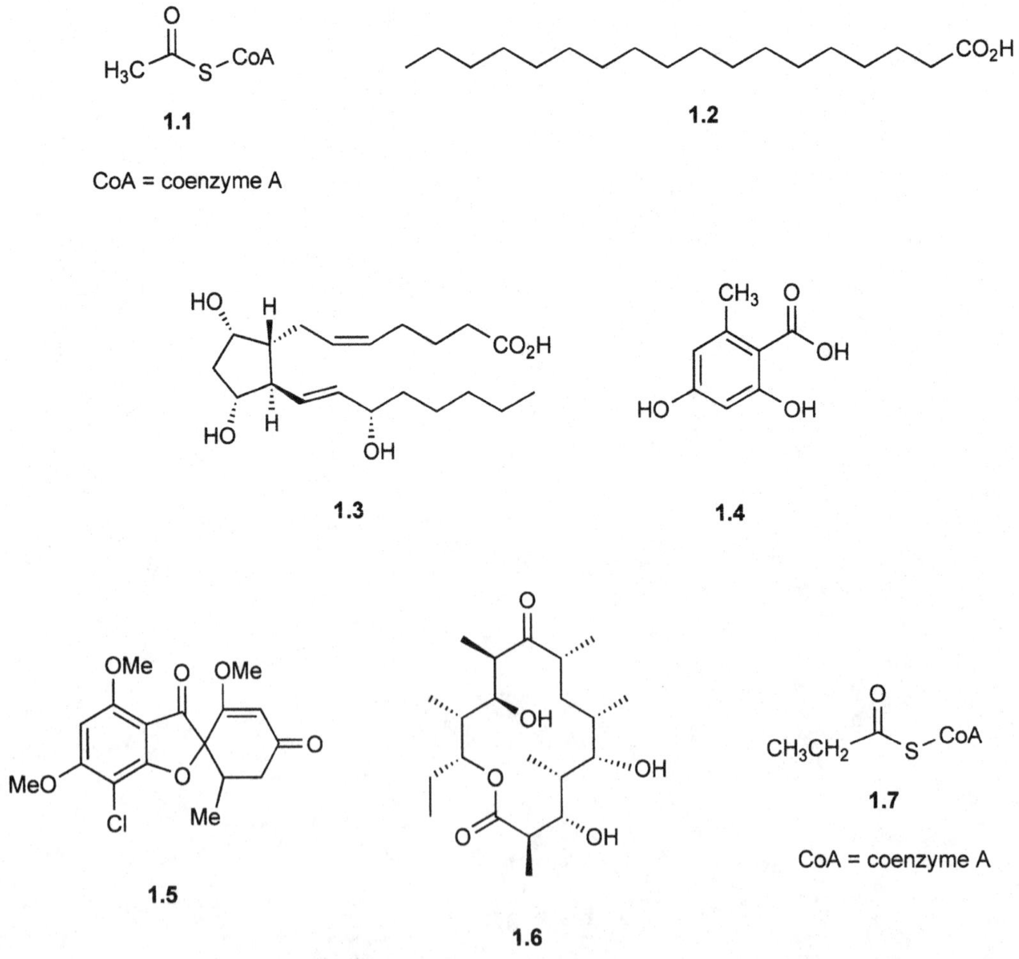

CoA = coenzyme A

Figure 1.1 Natural products derived from acetyl coenzyme A **1.1** and propionyl coenzyme A **1.7**.

2

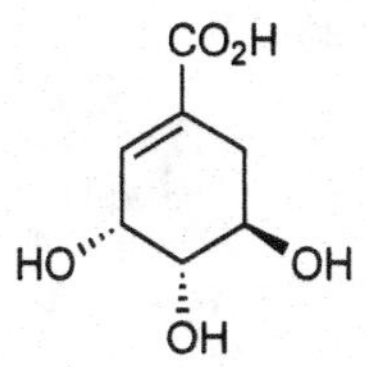

Figure 1.2 Natural products derived from shikimic acid **1.8**.

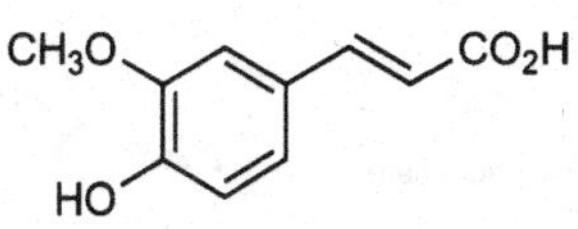

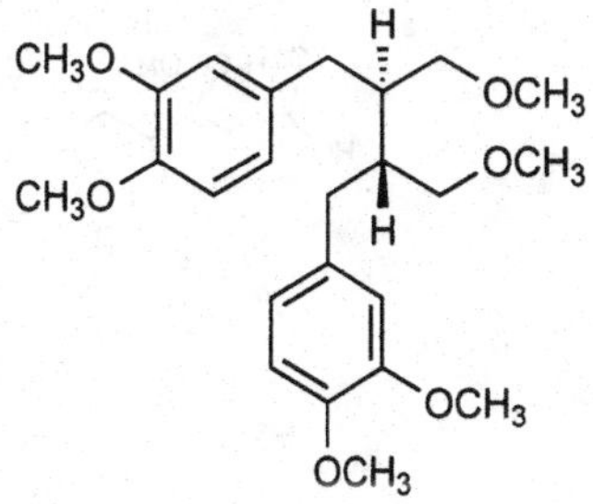

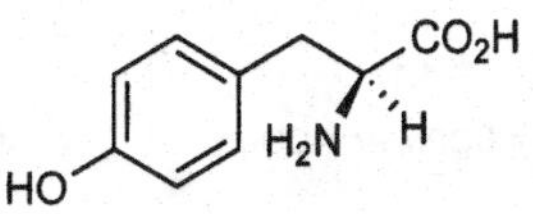

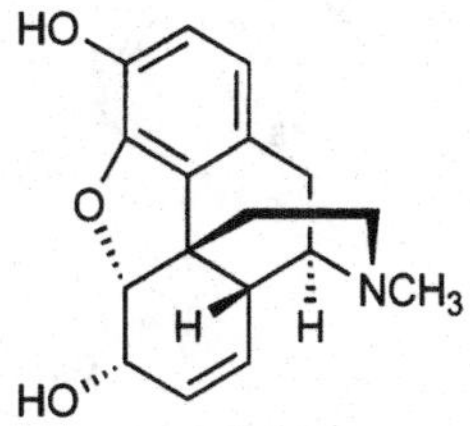

products. A representative example is 6-deoxyerythronolide B **1.6**, a precursor to several macrocyclic antibiotics, which is constructed from propionyl coenzyme A **1.7**. The underlying chemistry in these three sections is carbanion chemistry and the analogy between traditional organic reactions and their biological counterparts that proceed through carbanion intermediates is highlighted.

Shikimic acid **1.8** (Figure 1.2) is also produced from glucose in several steps and is the precursor to numerous aromatic compounds e.g. the phenolic compound ferulic acid **1.9** and phyllanthin **1.10**. Shikimic acid also produces aromatic amino acids, e.g. tyrosine **1.11**, from which many alkaloids such as morphine **1.12** are biosynthesized. Natural products derived from shikimic acid are considered in section 5. The underlying theme in this section is the chemistry of phenoxy radicals (PhO˙) and how these radicals can undergo dimerization to produce the framework of many natural products.

1.13

1.14

P = phosphate

1.15

1.16

1.17

Bz = COPh, Ac = COMe

1.18

1.19

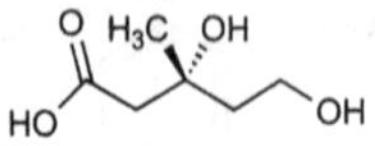

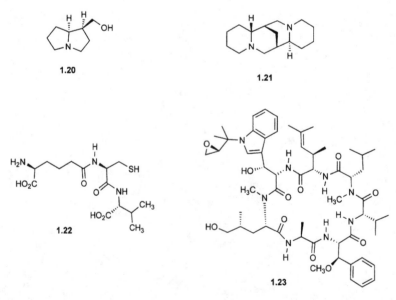

Figure 1.3 Terpene natural products.

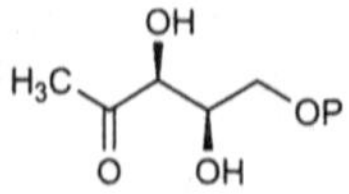

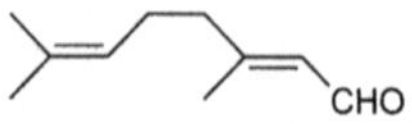

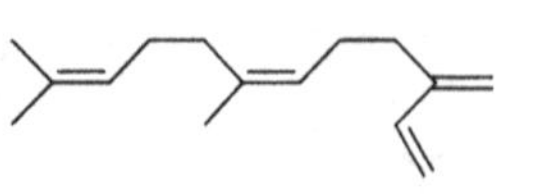

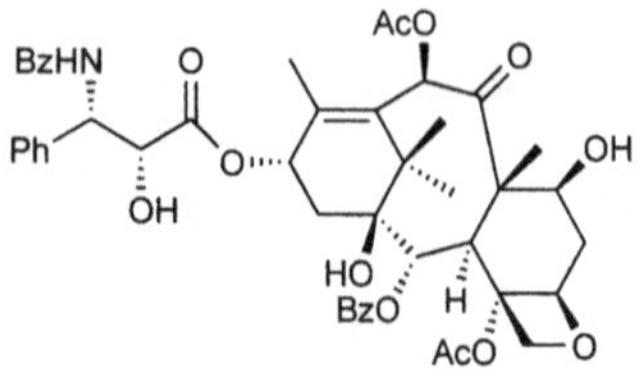

In section 6 the formation of mevalonic acid **1.13** is considered (Figure 1.3). Mevalonic acid is produced from two molecules of acetyl coenzyme A **1.1** and is one of two possible precursors to the terpene class of natural products. The other precursor to the terpene class of natural products is deoxyxylulose-5-phosphate **1.14** which is produced in several steps from glucose. Terpene natural products (or terpenoids) include relatively 'simple' molecules such as the fragrance *trans* citral **1.15** and the aphid repellent farnesene **1.16**, through to exotic structures as exemplified by the anti-cancer agent taxol **1.17**. Additionally, many other important types of natural products, including steroids such as cholesterol **1.18** and carotenoids such as vitamin A **1.19**, are terpenes. The underlying chemistry in terpene biosynthesis is the formation and reactions, particularly rearrangements, of carbocations.

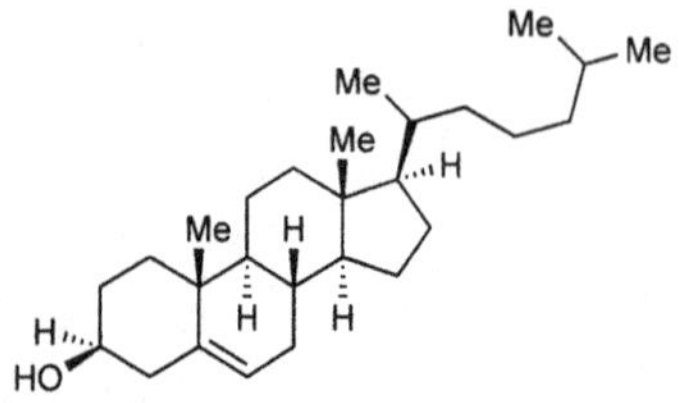

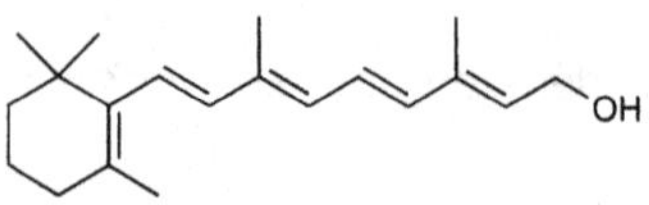

Many natural products are derived from amino acids and we have already noted above that aromatic amino acids are produced from shikimic acid **1.8**. Section 7 gives an overview of some natural products that are biosynthesized from aliphatic amino acids. Examples include the alkaloid laburnine **1.20** and sparteine **1.21**, the tripeptide **1.22** which is the precursor of many penicillins and the cyclic peptide, cyclomarin A **1.23** (Figure 1.4).

1.20

1.21

1.22

1.23

Figure 1.4 Natural products derived from amino acids.

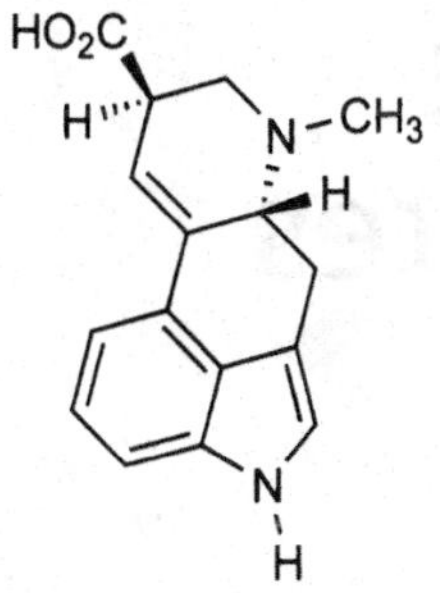

Figure 1.5 Natural products of mixed biosynthetic origin.

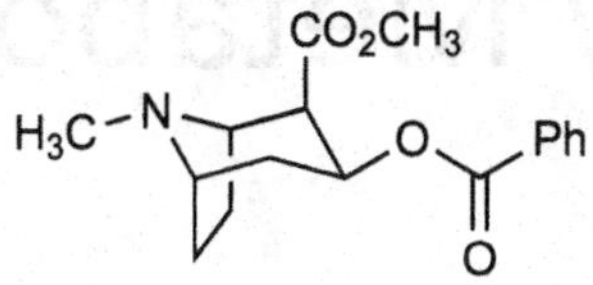

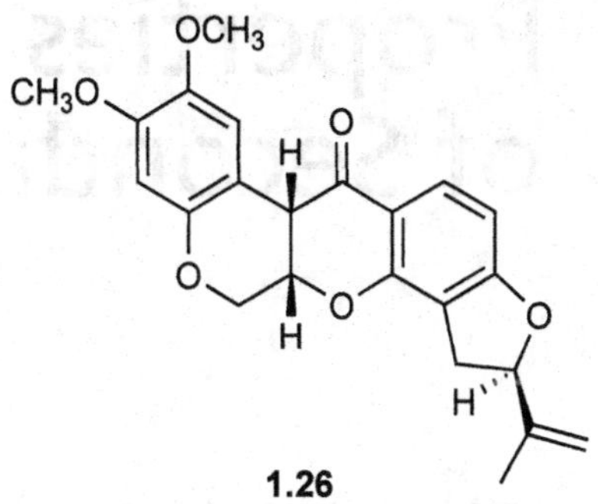

It is not always possible to categorize natural products according to a single key intermediate employed for their construction, because many natural products use two or even more basic building blocks. These compounds are said to be of mixed biosynthetic origin and examples include lysergic acid **1.24** (shikimic acid **1.8** and terpenoid origin), cocaine **1.25** (amino acid, acetyl coenzyme A **1.1** and shikimic acid **1.8** origin) and the insecticide rotenone **1.26** (acetyl coenzyme A **1.1**, shikimic acid **1.8** and terpenoid origin) (Figure 1.5). Natural products that are of mixed biosynthetic origin (including alkaloids which are often considered as a separate class of natural product) have been included at the most appropriate locations in this book.

In summary, the basic building blocks for secondary metabolites are:

- Acetyl coenzyme A **1.1** (precursor to fatty acids, polyketides).
- Shikimic acid **1.8** (precursor to aromatic compounds).
- Mevalonic acid **1.13** and deoxyxylulose-5-phosphate **1.14** (precursors to terpenoids).
- Amino acids (precursors to peptides, some alkaloids, penicillins).

1.2 Properties and Purpose of Secondary Metabolites

It has been suggested that secondary metabolic pathways have evolved as a means of consuming excess primary metabolites i.e. acetate, shikimic acid and amino acids. Many secondary metabolites are certainly not *waste products* but have been shown to regulate interactions between organisms. Examples include sex pheromones such as compounds **1.27** and **1.28** (Figure 1.6) which are produced by many species of insects to regulate their reproduction, and the production by many types of plants of substances to repel insect attack. The secondary metabolite-mediated interactions between organisms is the domain of 'chemical ecology' and there are several accounts of this area in the further reading list.

Many secondary metabolites have interesting pharmacological properties. For centuries herbs have been used by many civilizations for treatment of a variety of illnesses, and secondary metabolites often provide the therapeutically active component. Readers who are interested in the medicinal aspects of secondary metabolites should consult the accounts in the further reading list.

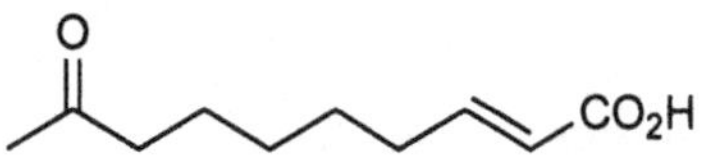

1.27

honeybee pheromone

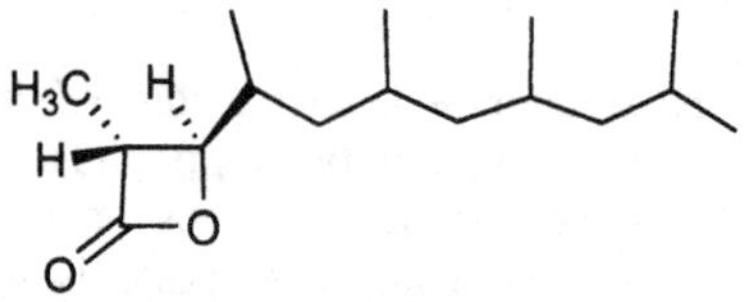

1.28

striped cucumber beetle pheromone

Figure 1.6   Pheromones.

Acetyl Coenzyme A: A Key Biological Intermediate

Objectives

After studying this section you will be able to:

(a) Recognize the similarity between acetyl coenzyme A reactions and organic reactions.
(b) Describe how malonyl coenzyme A is formed from acetyl coenzyme A.
(c) Identify the roles of acetyl coenzyme A and malonyl coenzyme A in biosynthesis.

2.1 What is Acetyl Coenzyme A?

Acetyl coenzyme A is a *thioester* derivative of acetic acid **2.1** (Figure 2.1). Acetic acid is a typical example of a *carboxylic acid* and methyl acetate **2.2** is a typical example of an *ester* derivative of acetic acid. Thioesters are simply the sulphur analogues of esters and compound **2.3** is a thioester analogue of compound **2.2**. Acetyl coenzyme A is also a thioester and for the moment we will represent it by the abbreviated thioester structure **2.4** which is the usual way of drawing this molecule. The full structure of the SCoA component of acetyl coenzyme A is given in Figure 2.2; you need only to recognize that acetyl coenzyme A is a thioester derivative. As we shall see in section 2.3, acetyl coenzyme A provides a *two-carbon fragment* (labelled *1* and *2* in structure **2.4**) in biosynthesis. This two-carbon atom fragment consisting of a methyl group and a carbonyl group is called an *acetyl group*, hence the name acetyl coenzyme A for structure **2.4**. Acetyl coenzyme A is usually referred to as a biological equivalent of *acetate* ($CH_3CO_2^-$) because it can provide a two-carbon acetyl fragment in biosynthetic reactions.

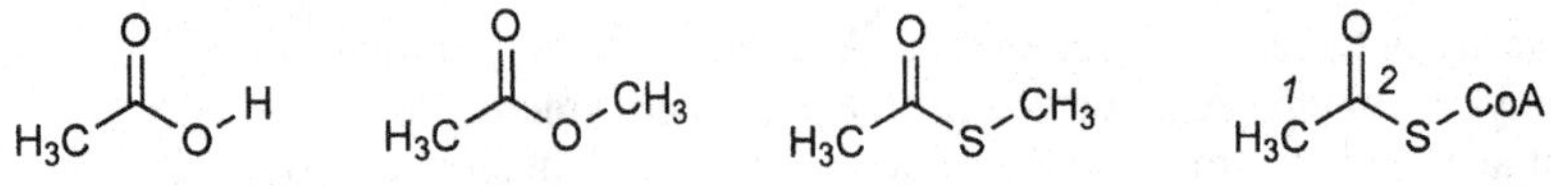

2.1   **2.2**   **2.3**   **2.4**

Figure 2.1 Esters and thioesters.

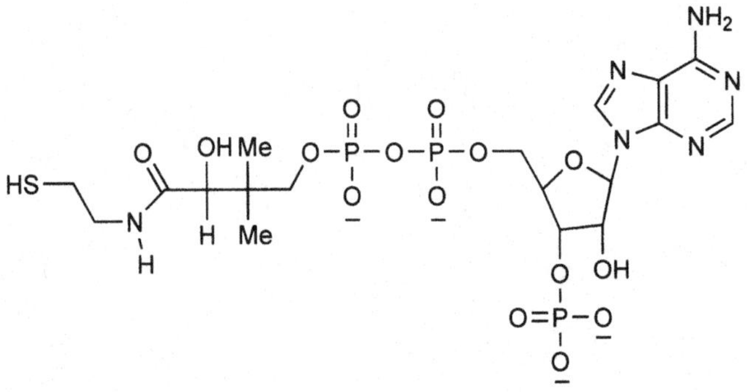

Figure 2.2 Coenzyme A.

2.2 Comparison of Organic and Acetyl Coenzyme A Reactions

Before we consider how acetyl coenzyme A is used in biosynthesis it is useful to consider some related organic reactions. Figure 2.3 shows the reaction between the acid chloride **2.5** derivative of acetic acid **2.1** and a typical *nucleophile* such as methanol. In this reaction, the nucleophile attacks at the carbonyl group of compound **2.5** giving the reaction intermediate **2.6** which then expels the good *leaving group* (Cl⁻) giving the product, methyl acetate **2.2**. The reaction shown in Figure 2.3 is a typical organic reaction called nucleophilic acyl substitution, because the —Cl group is replaced by the —OMe group. In principle, carboxylic acids such as acetic acid could undergo nucleophilic acyl substitution, but this type of reaction is generally inefficient because hydroxide is a poor leaving group. The conversion of carboxylic acids into their more reactive acid chlorides such as compound **2.5** is therefore carried out first because the acid chlorides are much more reactive towards nucleophiles. Thus, one way to go from acetic acid to a product **2.2** is to *activate* the carboxylic acid group by converting it into the acid chloride **2.5**. It must be emphasized that there are many ways of activating carboxylic acids in organic chemistry and formation of acid chlorides is just one representative method of doing this. In the same way that acetic acid can be activated towards nucleophilic substitution by converting it into its acid chloride it can also be activated in biological systems as acetyl coenzyme A **2.4**; the coenzyme A substituent is a good leaving group.

A second type of organic reaction we need to consider is *deprotonation*. The methyl protons of esters such as **2.2** are relatively acidic and can be removed by the action of a suitable base (Figure 2.4). The reason for the relative acidity of the methyl protons is that the resulting negative charge on the carbanion can be *delocalized* over both the methyl carbon atom and the carbonyl oxygen atom as shown in Figure 2.4. In a similar way, acetyl coenzyme A can be deprotonated in biological systems by a suitably basic group of an enzyme.

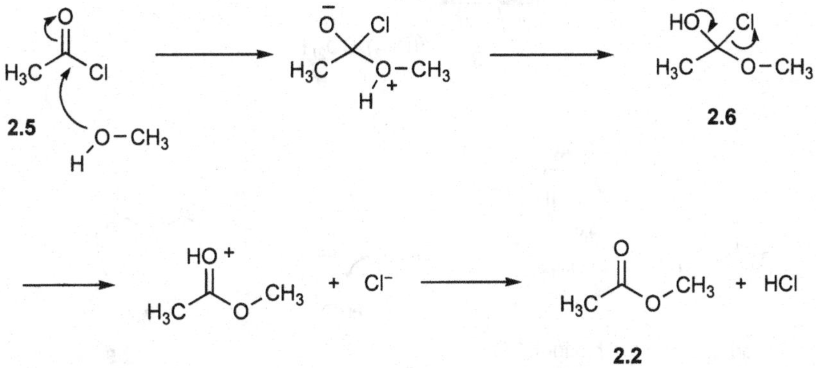

Figure 2.3 Nucleophilic substitution.

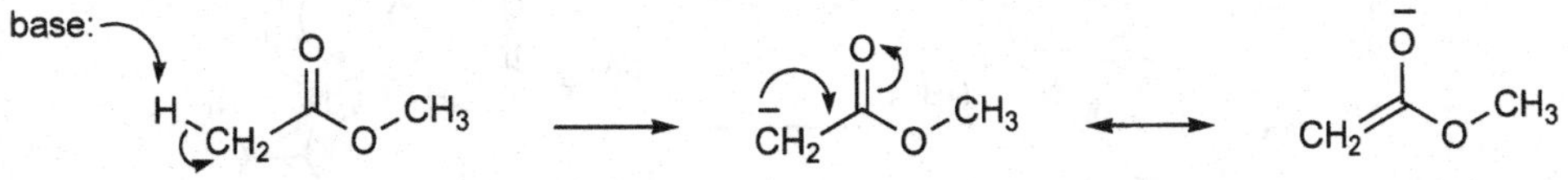

Figure 2.4 Ester deprotonation.

2.3 Malonyl Coenzyme A – A Partnership with Acetyl Coenzyme A

Acetyl coenzyme A works closely together with another molecule, *malonyl coenzyme A* **2.7** (Figure 2.5) in biosynthetic reactions. Malonyl coenzyme A is a thioester derivative of the dicarboxylic acid, *malonic acid* **2.8**. Malonyl coenzyme A is formed from acetyl coenzyme A by *carboxylation*, a reaction that requires the cofactor *biotin*. The full structure of biotin is shown in Figure 2.6. Biotin is attached through its carboxyl group to an associated enzyme, biotin carboxyl carrier protein (BCCP), during the carboxylation reaction of acetyl coenzyme A. The mechanism of the carboxylation reaction of acetyl coenzyme A is shown in Figures 2.7 and 2.8. The initial reaction involves the carboxylation of the biotin–BCCP complex with bicarbonate giving molecule **2.9**, as shown in Figure 2.7.

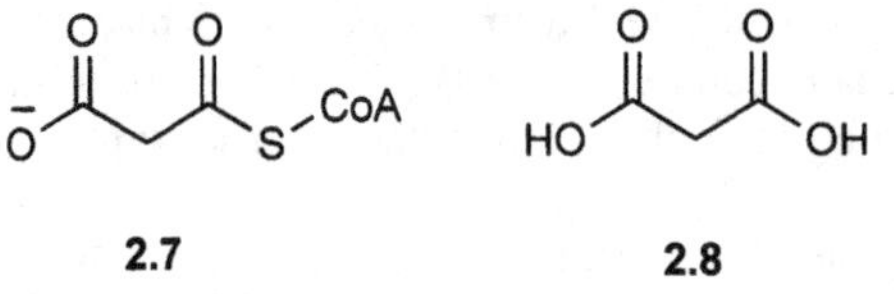

Figure 2.5   Malonyl coenzyme A **2.7** and malonic acid **2.8**.

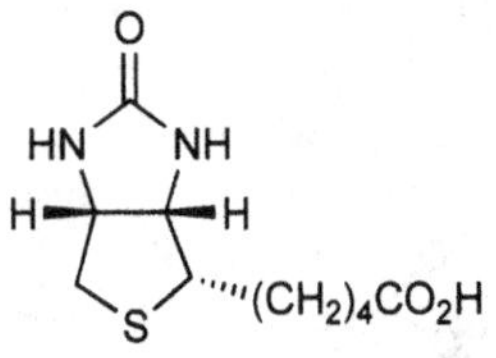

Figure 2.6   Biotin.

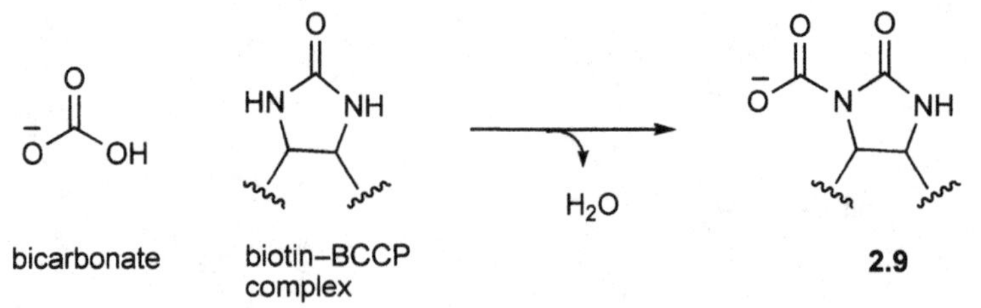

Figure 2.7   Carboxylation of biotin–BCCP complex.

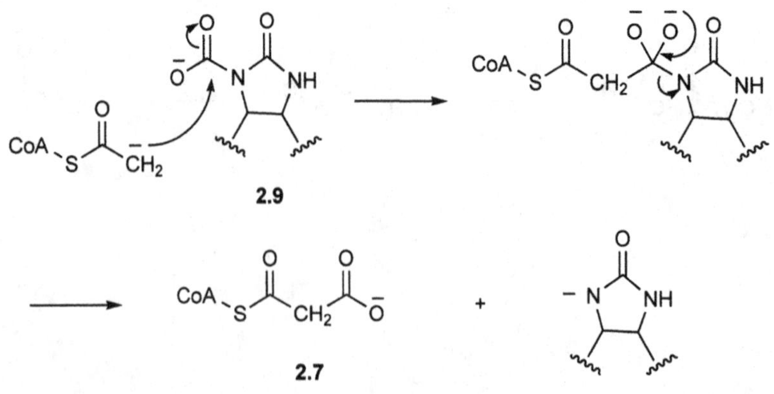

Figure 2.8   Formation of malonyl coenzyme A **2.7**.

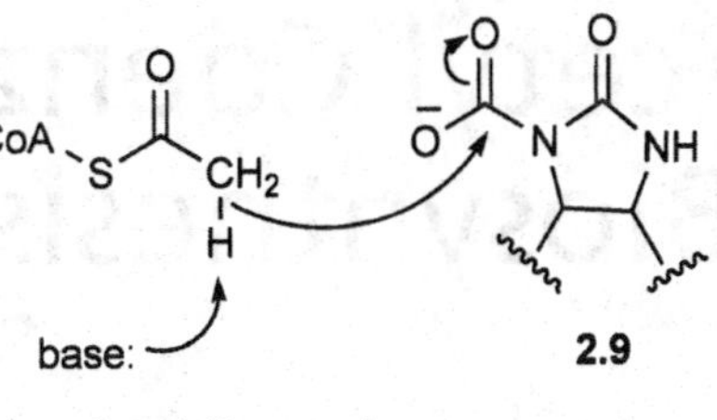

Figure 2.9 Synchronous deprotonation–nucleophilic attack.

The carboxylation of acetyl coenzyme A **2.4** by molecule **2.9** can be represented as a nucleophilic attack of the carbanion of acetyl coenzyme A, as shown in Figure 2.8. We have already seen above that acetyl coenzyme A can be deprotonated giving a carbanion derivative. The reaction shown in Figure 2.8 is also an example of a nucleophilic acyl substitution reaction which is mechanistically similar to the one shown in Figure 2.3. In Figure 2.8, the carbanion derivative of acetyl coenzyme A is the nucleophile and it attacks at the carbonyl group of the biotin–BCCP complex **2.9** as shown. The anion of the biotin–BCCP complex is eventually expelled as the leaving group. The mechanism shown in Figure 2.8 is in fact a simplified representation of what actually happens. The carbanion of acetyl coenzyme A does not exist as a discrete intermediate, but as deprotonation commences there is a synchronous movement of electrons as shown in Figure 2.9. You will appreciate that this reaction is controlled by an enzyme, so the base, acetyl coenzyme A and the carboxylated biotin can all be brought together by the enzyme so that a synchronous reaction can occur.

2.4 How is Acetyl Coenzyme A Used in Biosynthesis?

Before acetyl coenzyme A and malonyl coenzyme A can be used in biosynthetic reactions, they are firstly attached to a multifunctional enzyme complex, the fatty acid synthetase; this is represented diagrammatically in Figure 2.10. The acetyl coenzyme A **2.4** has been drawn as a *stick* formula for simplicity, i.e. the methyl group has not been drawn out as CH_3. In this reaction the —SCoA groups of molecules **2.4** and **2.7** are both substituted by the thiol groups (—SH) of the enzyme complex, giving the complex **2.10**. The thiol groups are behaving as nucleophiles and the SCoA group is a leaving group in a nucleophilic acyl substitution reaction, an analogous reaction to that shown in Figure 2.3. Once the compounds **2.4** and **2.7** have been anchored to the fatty acid synthetase, a carbon–carbon bond formation reaction then occurs as shown in Figure 2.11. This reaction can be viewed as a decarboxylation reaction giving a carbanion (which you will notice can be stabilized by delocalization of the negative charge), followed by a familiar nucleophilic substitution. You will notice that we have now synthesized a four-carbon atom chain, the atoms labelled *a* originated from acetyl coenzyme A and the atoms labelled *m* originated from malonyl coenzyme A. Since malonyl coenzyme A came originally

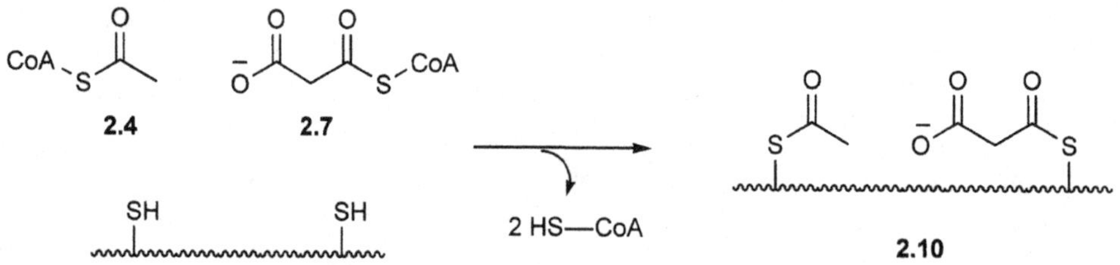

Figure 2.10 Fatty acid synthetase complex **2.10**.

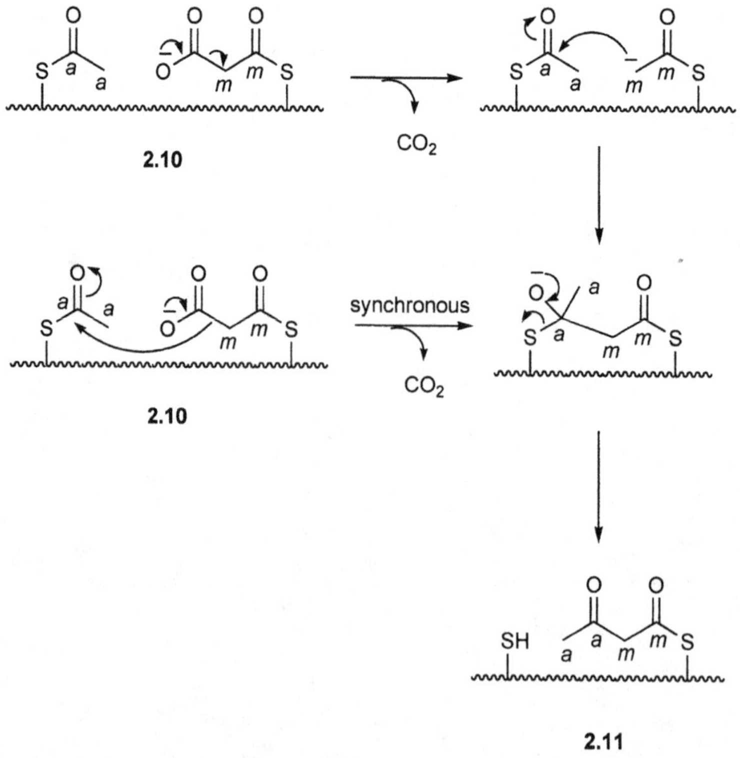

Figure 2.11 Carbon–carbon bond formation.

from acetyl coenzyme A, we have, in effect, condensed two molecules of acetyl coenzyme A to give a four-carbon atom unit **2.11**. In Figure 2.11, the reaction mechanism has been shown as a stepwise reaction involving the formation of a discrete carbanion intermediate. In reality, a synchronous reaction occurs; as carbon dioxide is lost carbon–carbon bond formation takes place as shown in Figure 2.11. Similarly, analogues of either acetyl coenzyme A or malonyl coenzyme A can be condensed. Figure 2.12 illustrates the reaction of propionyl coenzyme A **2.12** with malonyl coenzyme A which gives the intermediate **2.13**. In structure **2.13**, the atoms labelled p originated from proplonyl coenzyme A.

Intermediates **2.11** and related structures are used widely as precursors to a diverse range of natural products. In sections 3 and 4 we shall see how fatty acids and polyketides are biosynthesized from intermediates **2.11** and related structures.

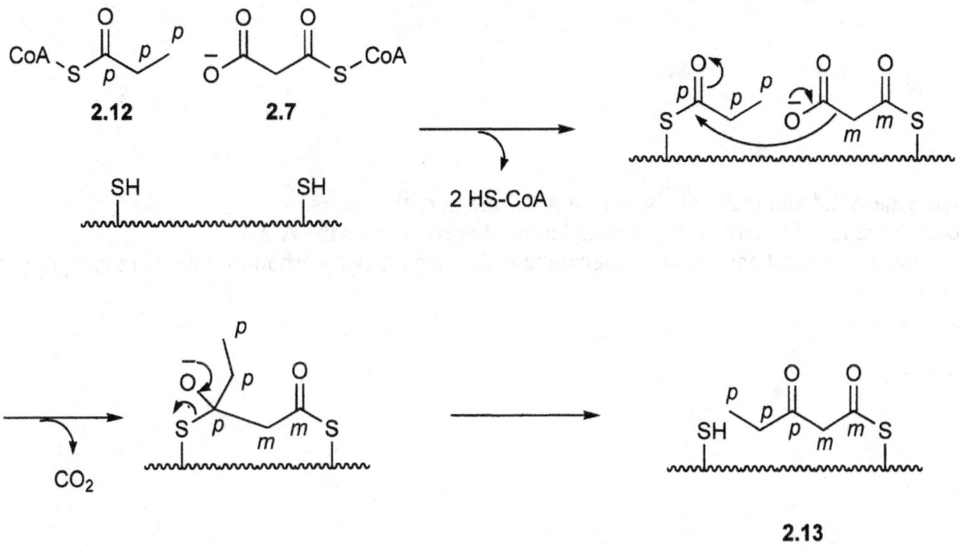

Figure 2.12   Formation of complex **2.13**.

Summary

(a) Acetyl coenzyme A **2.4** and malonyl coenzyme A **2.7** are both thioesters.
(b) Malonyl coenzyme A **2.7** is formed by carboxylation of acetyl coenzyme A **2.4**.
(c) Both acetyl coenzyme A **2.4** and malonyl coenzyme A **2.7** provide two-carbon fragments in biosynthetic reactions.

Biosynthesis of Fatty Acids

Objectives

After studying this section you will be able to:

(a) Recognize the structure and describe the function of fatty acids in biological systems.
(b) Describe how even-numbered, odd-numbered and branched saturated fatty acids are biosynthesized.
(c) Describe how unsaturated fatty acids are produced from saturated fatty acids.
(d) Describe how fatty acids can undergo further structural modifications.

3.1 What are Fatty Acids?

Fatty acids are *alkanoic acids* and the majority of naturally occurring fatty acids have straight-chains possessing an *even number* of carbon atoms. A typical example is octadecanoic acid, which has the trivial name stearic acid **3.1** (Figure 3.1). The most commonly found fatty acids consist of 10 to 20 carbon atoms with palmitic acid (16 carbon atoms) and stearic acid (18 carbon atoms) being particularly widespread. The fatty acids described above are examples of *saturated* fatty acids, but *unsaturated* fatty acids with an even number of carbon atoms are also commonly encountered. A typical example of an unsaturated fatty acid is 9-octadecenoic acid (Figure 3.1) which has the trivial name oleic acid **3.2**. Note that mono unsaturated fatty acids generally possess the *cis* geometry about the double bond.

Saturated fatty acids containing an *odd number* of carbon atoms are also found in Nature, but these compounds are not as common as their even-numbered counterparts. Other unusual types of fatty acids (Figure 3.2) are acetylenic acids e.g. crepenynic acid **3.3**, branched saturated fatty acids e.g. tuberculostearic acid **3.4** and branched unsaturated fatty acids e.g. sterculic acid **3.5**, which also possesses a cyclopropene ring.

A shorthand representation of the structures of fatty acids has been developed which takes into account chain length, the position and stereochemistry (*cis* or *trans*) of any double bonds and the position of any acetylenic bonds. Thus, stearic acid **3.1**, oleic acid **3.2** and crepenynic acid **3.3** have the abbreviations:

- **3.1** – 18:0 (18-carbons: 0 points of unsaturation)
- **3.2** – 18:1 (9c) [18-carbons: 1 point of unsaturation (*cis* double bond at the 9-position)
- **3.3** – 18:2 (9c, 12a) [18-carbons: 2 points of unsaturation (*cis* double bond at the 9-position, acetylenic bond at the 12-position)].

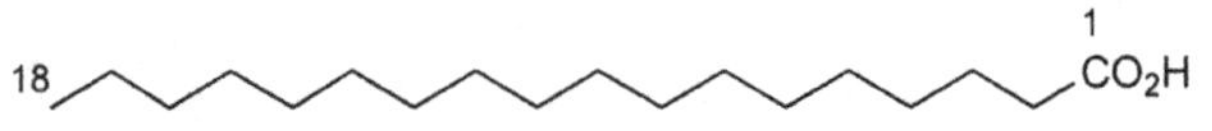

3.1

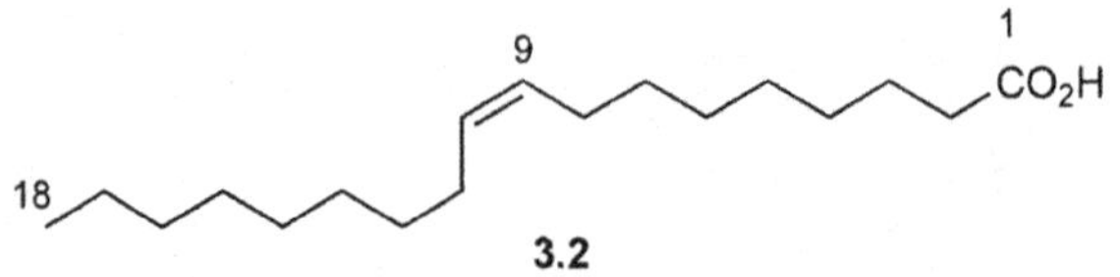

3.2

Figure 3.1   Stearic **3.1** and oleic **3.2** acids.

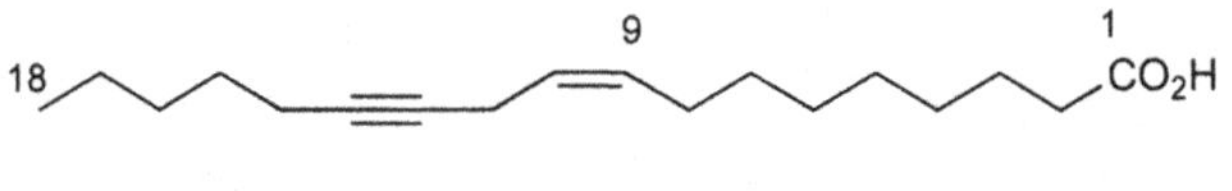

3.3

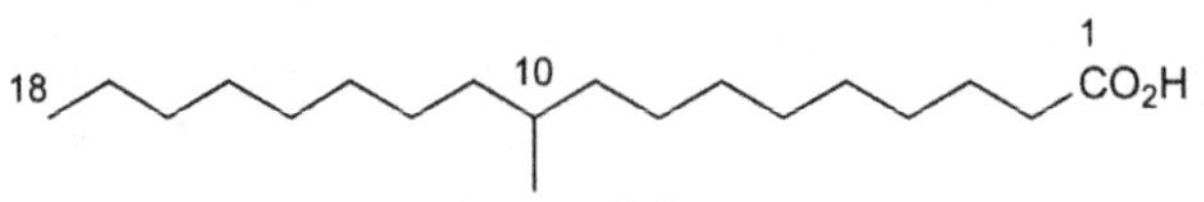

3.4

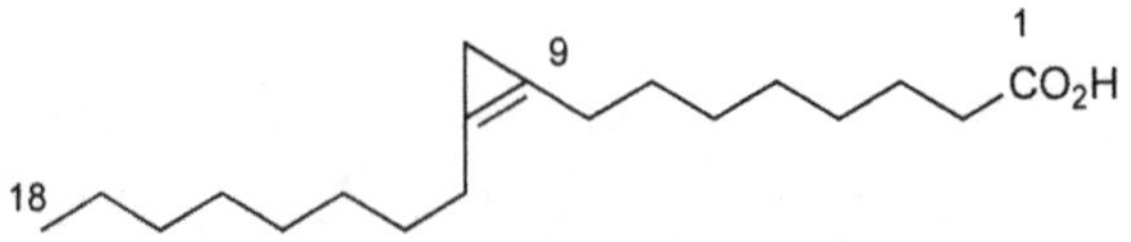

3.5

Figure 3.2   Acetylenic, branched and odd-numbered fatty acids.

Problem

Problem 3.1 What are the structures of arachidonic acid 20:4 (5c, 8c, 11c, 14c) and dehydromatricaria acid 10:4 (2t, 4a, 6a, 8a)?

3.2 Occurrence and Function of Fatty Acids

Fatty acids are usually encountered in Nature as their ester derivatives and these ester derivatives are collectively known as *lipids*, a term which recognizes their insolubility in water. For example, the *mono-ester* derivatives **3.6** (Figure 3.3) formed from long chain fatty acids (R^1CO_2H) and long chain alcohols (R^2OH) are termed *waxes* whose biological function is to provide protective coatings on skin, fur, leaves etc. Waxes are also storage products in some plants and animals. Tri-ester derivatives of fatty acids are extremely important biological molecules and these compounds are formed from the alcohol, *glycerol* **3.7** (which possesses three hydroxyl groups) and three fatty acids. Thus, each hydroxyl group from glycerol forms an ester linkage with a fatty acid, giving a tri-ester **3.8**. Molecules such as **3.8** are termed *glycerides* (or acylglycerols) and these compounds are the constituents of *fats* and *oils*. Glycerides act as an energy store in biological systems; they can be broken down into their constituent fatty acids which can subsequently undergo a process termed *β-oxidation* (see section 3.8). In *β-oxidation*, the fatty acid is broken down, two carbon atoms at a time, into acetyl coenzyme A units to provide energy. Thus, stearic acid would yield 9 molecules of acetyl coenzyme A if it were completely oxidized. Molecules related to tri-esters **3.8** such as the phosphoglyceride **3.9** are important constituents of cell membranes.

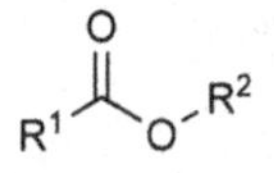

3.6

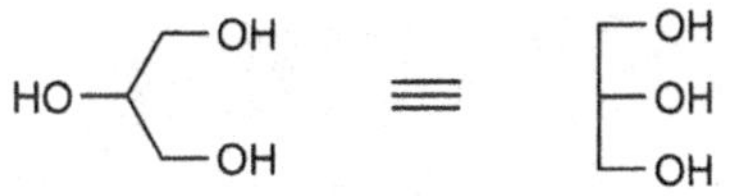

R^1 = alkyl chain of a fatty acid
R^2 = alkyl chain of an alcohol

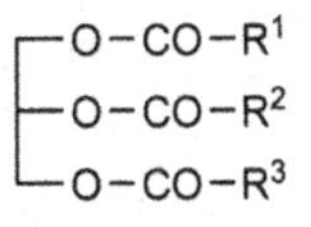

3.7

3.8

R^1, R^2 and R^3 =
alkyl chain of a fatty acid

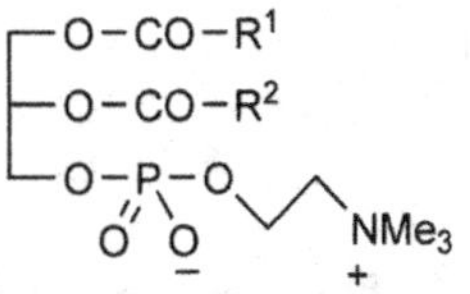

3.9

R^1 and R^2 =
alkyl chain of a fatty acid

Figure 3.3 Ester derivatives of fatty acids.

Problem

Problem 3.2 What are the possible structures of the glyceride formed from two molecules of stearic acid **3.1** and one molecule of oleic acid **3.2**?

3.3 Biosynthesis of Saturated Straight-Chain Fatty Acids

The starting point for fatty acid synthesis is the fatty acid synthetase complex that we have already met in section 2 (see Figure 2.11, structure **2.11**). This complex is redrawn in an abbreviated form **3.10** in Figure 3.4 where Enz represents the enzyme component of the complex. Remember that, in structure **3.10**, the carbon atoms labelled *a* are derived from acetyl coenzyme A and the carbon atoms labelled *m* are derived from malonyl coenzyme A. Thus, acetyl coenzyme A provides the *starter* group. The ketone carbonyl group in structure **3.10** is then reduced to the alcohol **3.11** by enzymes known as *dehydrogenases* working together with a *cofactor*, nicotinamide adenine dinucleotide phosphate *NADPH*. You should appreciate that NADPH acts as a nucleophilic source of *hydride* (H^-) and the H in the abbreviation NADPH reflects this. NADPH is a 1,4-dihydropyridine derivative and the mechanism by which NADPH provides hydride is shown in Figure 3.5. The compound NADPH is therefore similar to the hydride-based reducing agents such as sodium borohydride ($NaBH_4$) which are commonly used to effect reductions in organic chemistry. Both NADPH and $NaBH_4$ can be considered as nucleophilic sources of hydride. Note that the two hydrogen atoms at the 4-position of the pyridine ring in NADPH are non-equivalent. This can be appreciated by analogy with your left hand – if you place your left hand on a surface with your thumb representing the amide ($-CONH_2$) group you can distinguish the palm and the back of

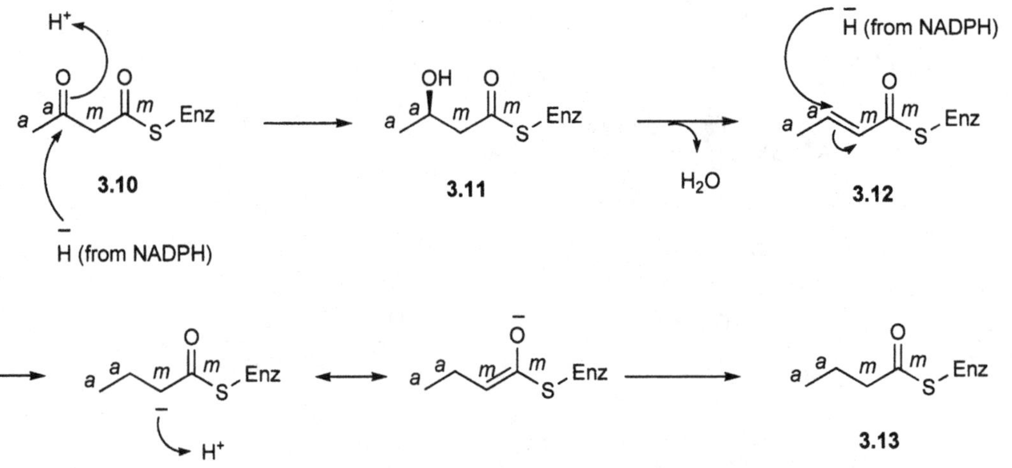

Figure 3.4 Ketone reduction.

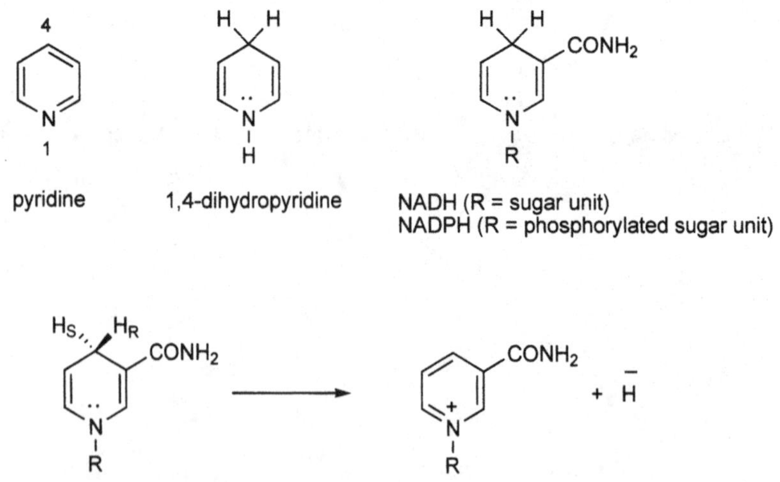

pyridine   1,4-dihydropyridine   NADH (R = sugar unit)
NADPH (R = phosphorylated sugar unit)

Figure 3.5 1,4-Dihydropyridines and NADPH.

your hand. Similarly the hydrogen atom at the 'palm' of NADPH can be distinguished from the hydrogen at the *back*. These hydrogen atoms have been labelled as H_R and H_S. The mechanism of the reduction of ketone **3.10** giving alcohol **3.11** can therefore be represented as involving a nucleophilic attack of hydride at the carbonyl group, as shown in Figure 3.4. In this stereospecific reduction, the hydrogen H_S is transferred from the NADPH giving the alcohol **3.11** with the *R*-stereochemistry. The alcohol **3.11** is then dehydrated giving the alkene **3.12** which is then reduced by NADPH (H_R is used in this reduction) forming the product **3.13**. The overall result of the reaction sequence shown in Figure 3.4 is reduction of the ketone carbonyl group ($>C{=}O$) to methylene ($>CH_2$).

Note that in the reduction of compound **3.12** giving compound **3.13** (Figure 3.4), hydride is being added to the alkene fragment rather than the carbonyl group of compound **3.12**. The reason why this type of nucleophilic addition (usually referred to as *conjugate addition* or *Michael addition*) takes place is because the negative charge in the intermediate anion can be stabilized by delocalization over the carbonyl group.

Problem

Problem 3.3 Most elimination reactions in organic chemistry are *anti* eliminations rather than *syn* eliminations, as shown below. The labelling studies indicated below have enabled the stereochemistry of the dehydration reaction shown in Figure 3.4 to be determined. Is this dehydration a *syn* or *anti* elimination?

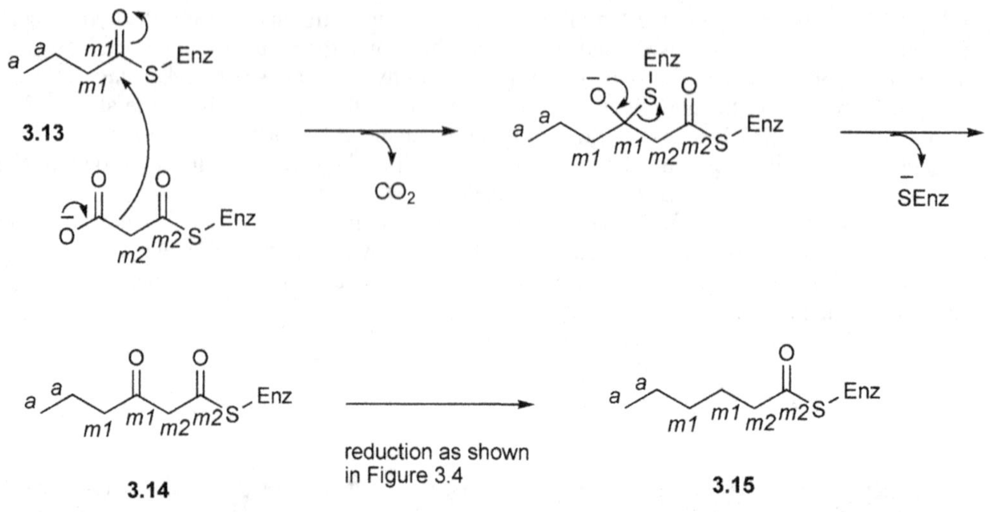

Figure 3.6 Addition of a malonyl unit.

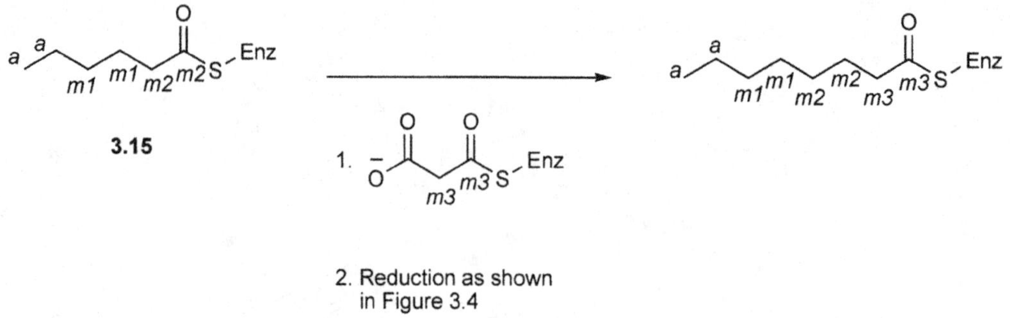

Figure 3.7 Chain elongation.

Once structure **3.13** has been formed it is then reacted with an enzyme-bound malonyl unit as shown in Figure 3.6, giving the ketone **3.14**. This reaction is **directly analogous** to that shown in Figure 2.11 of section 2. The ketone group in structure **3.14** is then reduced by exactly the same process as shown in Figure 3.4, giving intermediate **3.15**. Note that structure **3.15** has been built up from one acetyl coenzyme A precursor (labelled *a*) and two malonyl coenzyme A precursors (labelled *m1* and *m2*). The sequence in which structure **3.15** has been biosynthesized is therefore: $a \rightarrow a + m1 \rightarrow a + m1 + m2$. Further reaction of structure **3.15** with another malonyl coenzyme A unit (Figure 3.7) will then add a further two carbon atoms (labelled *m3*) and so on until the required number of carbon atoms have been added. For example, stearic acid which has 18 carbon atoms, is formed from an acetyl starter unit (which provides carbon atoms 17 and 18) and 8 malonyl units (which provide carbon atoms 1 to 16). This is the reason why most common fatty acids have an even number of carbon atoms; they are biosynthesized using a two-carbon starter unit and elongated by a series of additions of two-carbon fragments. Once the carbon chain has been built up to the correct length, hydrolysis of the thioester gives the fatty acid and HS-Enz. Thioesters can also react with nucleophiles other than water: alcohols give ester derivatives and amines yield amides.

Problem

Problem 3.4 The NADPH of thioester can yield either aldehydes or alcohols. Show how the thioester **3.15** could be reduced to the aldehyde $CH_3(CH_2)_4CHO$ and also to the alcohol $CH_3(CH_2)_4CH_2OH$.

The biosynthesis of fatty acids that possess an odd number of carbon atoms is essentially similar to the biosynthesis of their even-numbered counterparts, except that propionyl coenzyme A (a three carbon atom unit which you have met in section 2) is used as a starter (Figure 3.8). After attachment of propionyl coenzyme A to the fatty acid synthetase complex (structure **3.16**), chain elongation then occurs using a malonyl unit giving structure **3.17**, by an analogous mechanism to that already shown in Figure 3.6. Reduction of the ketone group in structure **3.17** by the mechanism shown in Figure 3.4 then gives the thioester **3.18** which is derived from propionyl coenzyme A (atoms labelled *p*) and malonyl coenzyme A (atoms labelled *m*). Further elongation of compound **3.18** with additional malonyl units then occurs until the chain has been built up.

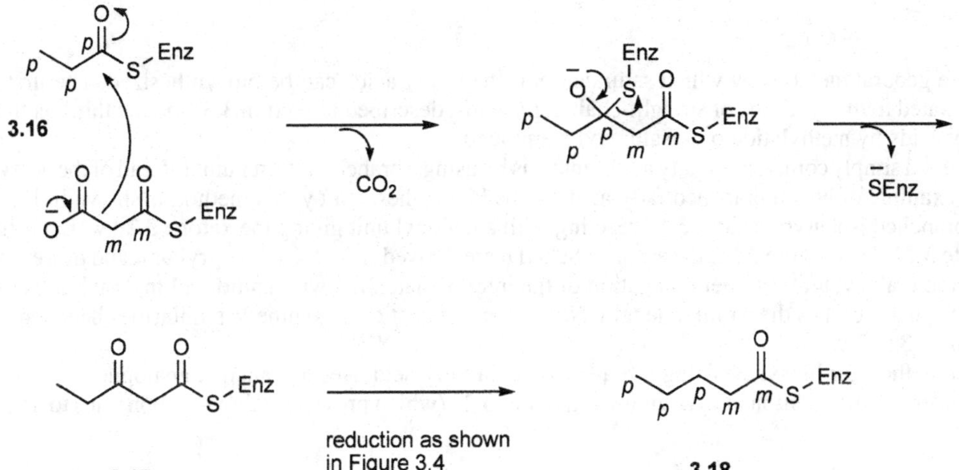

Figure 3.8 A propionyl starter unit.

3.4 Biosynthesis of Saturated Branched Fatty Acids

There are three general methods by which saturated branched fatty acids can be biosynthesized. The first and second methods are related to the synthesis of straight-chain fatty acids, described in section 3.3, and the third method produces branched fatty acids by methylation of unsaturated compounds.

The first method simply commences fatty acid synthesis by using a branched starter unit instead of the acetyl coenzyme A starter. An example of how a branched fatty acid can be biosynthesized by this method is shown in Figure 3.9, and involves the branched isobutyryl starter **3.19** reacting with a malonyl unit giving the ketone **3.20** which is then reduced to intermediate **3.21**. In structure **3.21**, the atoms labelled *b* are derived from the isobutyryl unit and those labelled *m* are derived from the malonyl unit. Further elongation of the intermediate **3.21** with additional malonyl units subsequently occurs until the chain reaches the required length. Note the mechanistic and synthetic similarities between the reactions shown in Figures 3.6–3.9.

The second method for biosynthesizing a branched chain fatty acid uses a branched analogue of the malonyl unit. Figure 3.10 illustrates the addition of a methylmalonyl unit **3.22** (which provides a three-carbon unit) to an acetyl starter

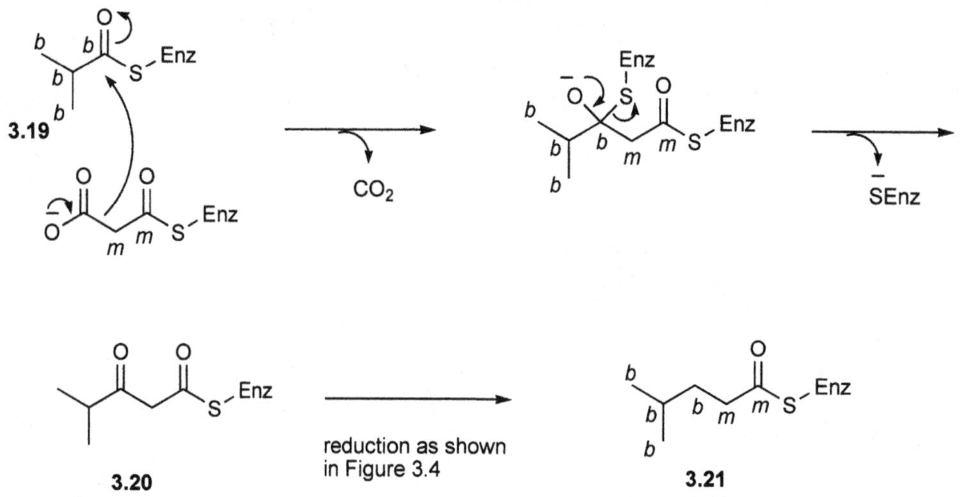

Figure 3.9 An isobutyryl starter unit.

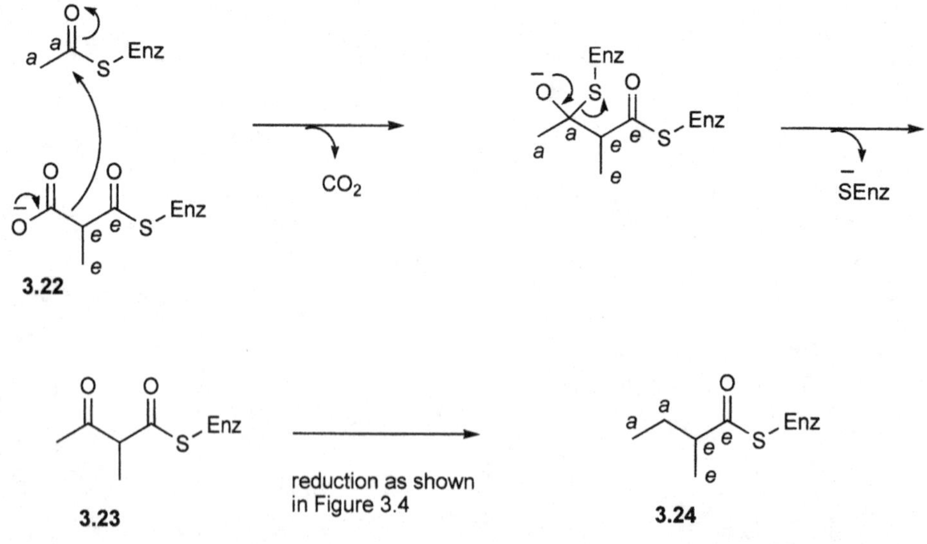

Figure 3.10 Branching using a methylmalonyl unit.

22

unit giving the ketone **3.23** which is then reduced to the intermediate **3.24**. In structure **3.24**, the atoms labelled *a* are derived from the acetyl unit and the atoms labelled *e* are derived from the methylmalonyl unit. Further elongation of structure **3.24** can then occur giving branched fatty acid derivatives.

Problem

Problem 3.5 Suggest a reasonable biosynthetic pathway to the fatty acid **3.25**.

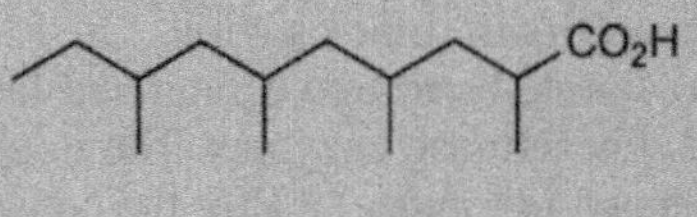

3.25

The third method by which branched fatty acids can be formed is by *methylation* of unsaturated fatty acids. This process is shown in Figure 3.11 and can provide both methylated and cyclopropane fatty acid derivatives. The biological methylating agent is a substance called S-*adenosyl methionine* (Figure 3.12). You should appreciate that S-adenosyl methionine has a sulphonium ($>SMe^+$) fragment which is responsible for its methylation properties, because the sulphide (R^1-S-R^2) is an excellent leaving group. Methylation of an unsaturated fatty acid therefore gives a *carbocation* intermediate **3.26**, which can then react with NADPH (arrow labelled *a*) giving a methylated product. Alternatively, the carbocation intermediate can lose a proton forming the cyclopropane ring (arrow labelled *b*).

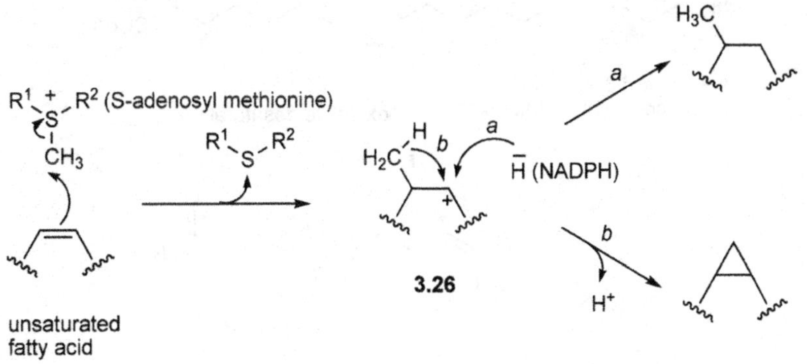

Figure 3.11 Methylation of unsaturated fatty acids.

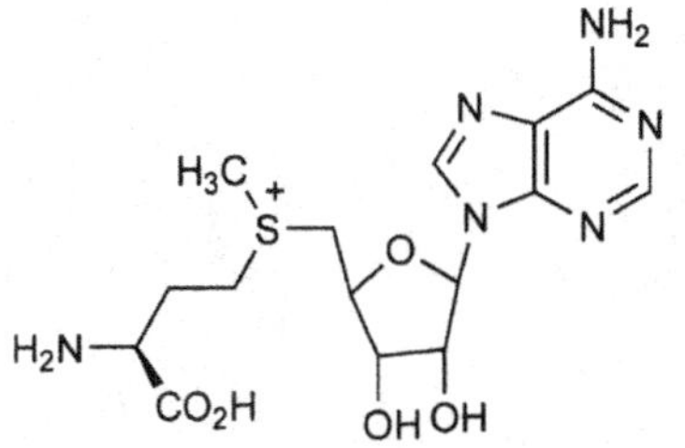

Figure 3.12 S-Adenosyl methionine.

3.5 Mono-unsaturated Fatty Acids

Unsaturated fatty acids can be produced by two general routes; the *aerobic* route which occurs in animals and plants, and the less common *anaerobic* route which occurs in some bacteria.

The aerobic route directly introduces the double bond into a saturated fatty acid precursor by a process known as *oxidative desaturation*, which is regulated by desaturase enzymes. The double bond is generally introduced between carbon atoms 9 and 10, and an example is shown in Figure 3.13 for the conversion of stearic acid into oleic acid. Note that the double bond has the *cis* configuration. Desaturase enzymes are given the prefix Δ^x where x refers to the position that the double bond is introduced; Figure 3.13 therefore involves a Δ^9 desaturase.

Oxidative desaturation appears to be closely linked with the oxidation of fatty acids (Figure 3.14). A reaction mechanism has been proposed which involves radical intermediates, and using iron as the oxidizing agent has been suggested. Note that in the desaturation reaction, the hydrogen atoms that are removed have a *syn* relationship.

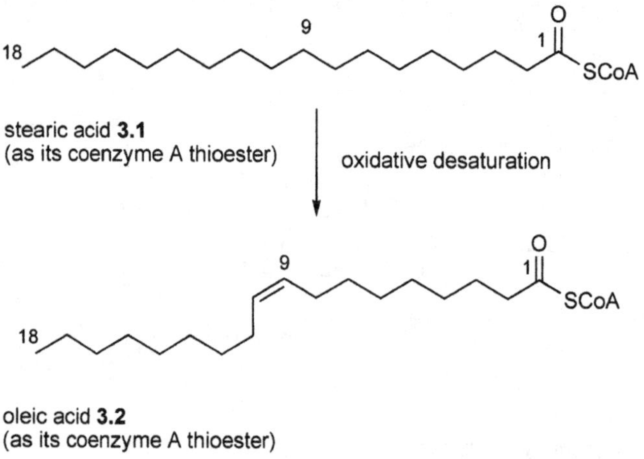

Figure 3.13 Oxidative desaturation of stearic acid **3.1**.

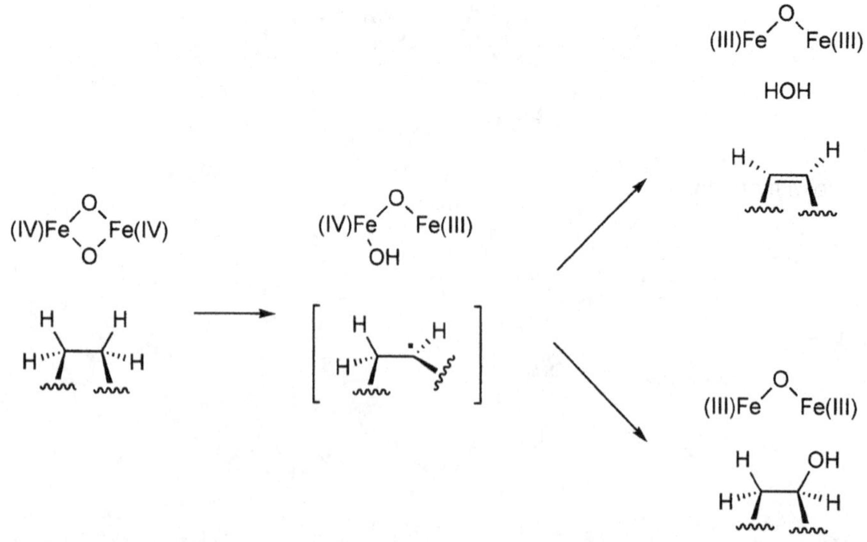

Figure 3.14 Scheme for oxidative desaturation.

Problems

Problem 3.6 What unsaturated product would you expect to obtain from oxidative desaturation of the acetyl coenzyme A thioester of palmitic acid, $CH_3(CH_2)_{14}CO_2H$?

Problem 3.7 How might tuberculostearic acid **3.4** and sterculic acid **3.5** be biosynthesized from stearic acid **3.1**?

The anaerobic route to oleic acid **3.2** is shown in Figure 3.15. In this route the structure **3.13** undergoes chain elongation by the biosynthetic pathways already discussed in Figures 3.4 and 3.7, giving the intermediate **3.27**. Structure **3.27** then reacts with a further malonyl unit giving ketone **3.28** which is then reduced to the alcohol **3.29**. Dehydration of the alcohol **3.29** gives the conjugated alkene **3.30**. Up to this point there is no difference between this biosynthetic pathway and the biosynthetic pathways to long-chain saturated fatty acids which have already been discussed. The *cis* double bond is introduced by isomerization of the conjugated alkene **3.30** giving the non-conjugated alkene **3.31**. This subsequently undergoes chain elongation by the process already described eventually giving oleic acid **3.2**. Note the shared common intermediate **3.30** in anaerobic biosynthesis of oleic acid, and its fully saturated analogue, stearic acid **3.1**. Also note that the reduction of compound **3.30** can occur by a conjugate addition mechanism (see Figure 3.4), whereas alkene **3.31** cannot be reduced by conjugate addition.

Additional points of unsaturation can be introduced by further oxidative desaturation leading to polyunsaturated fatty acids. It is interesting to note that double bonds (which have the *cis* configuration) usually occur in a *skipped* or *methylene interrupted* fashion, so that two double bonds are separated by a methylene ($>CH_2$) group. The 'skipped' double bonds are therefore *not conjugated*.

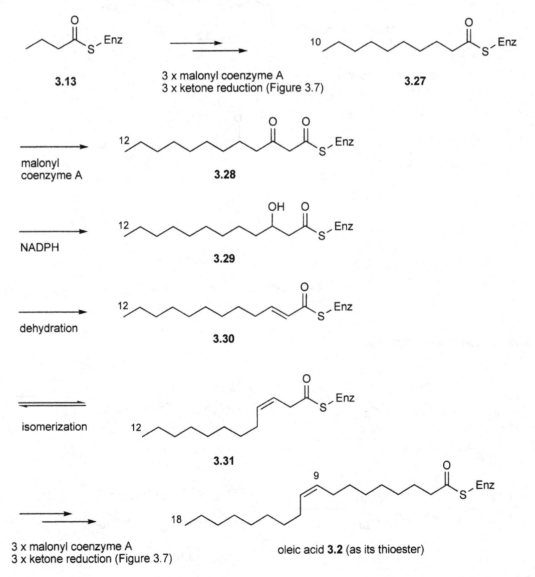

Figure 3.15 Anaerobic formation of oleic acid **3.2**.

3.6 Polyunsaturated Fatty Acids

The position in which new double bonds are introduced differs between animals and plants (Figure 3.16). Animals normally introduce new double bonds between *existing double bonds and the carboxyl group* of the fatty acid, as shown for the biosynthesis of the fatty acid **3.32** from oleic acid **3.2**. In contrast, plants usually introduce new double bonds between *existing double bonds and the methyl terminus* of the fatty acid, as depicted for the biosynthesis of linoleic acid **3.33** from oleic acid **3.2**.

Arachidonic acid **3.34** has a sequence of four *cis skipped* double bonds, and this substance is biosynthesized from linoleic acid **3.33** as shown in Figure 3.17. The transformation of linoleic acid into arachidonic acid involves biosynthetic conversions that you have already met in this section.

A 1,4-dehydrogenation process has also been identified in plants. The biosynthesis of calendic acid **3.35** from linoleic acid **3.33** (Figure 3.18) illustrates this process and this reaction is mediated by a $\Delta^{8,10}$ desaturase enzyme. Two hydrogen

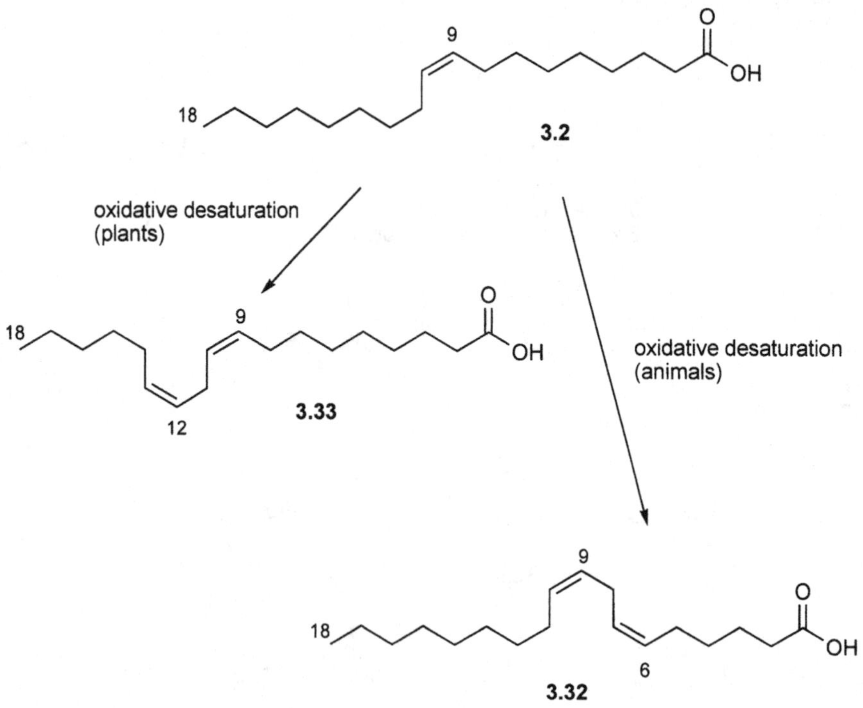

Figure 3.16 Polyunsaturated fatty acids.

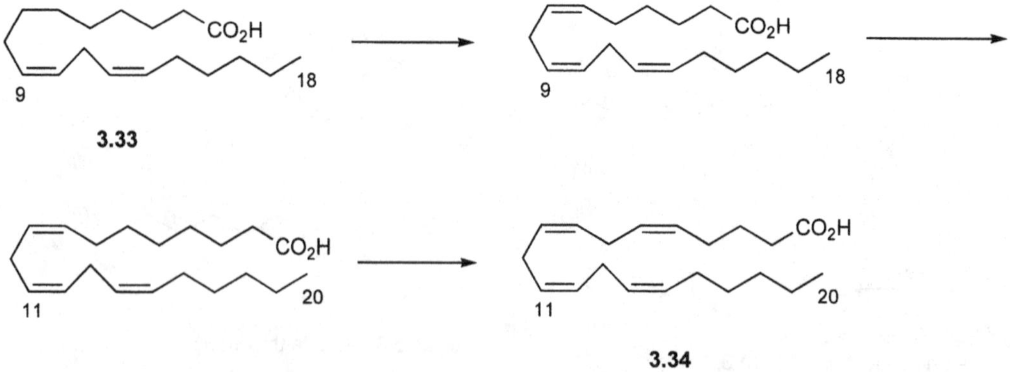

Figure 3.17 Arachidonic acid **3.34**.

26

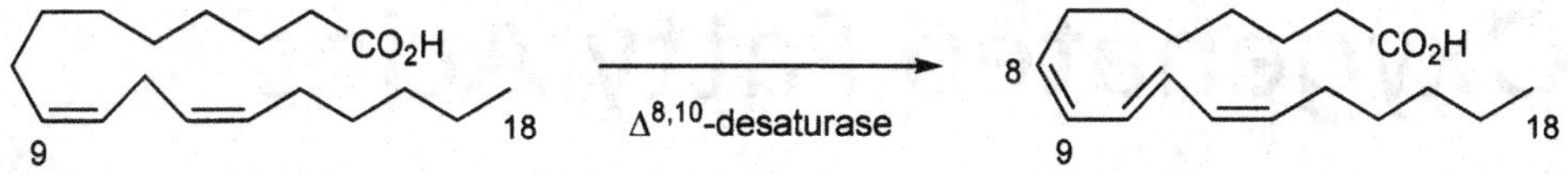

Figure 3.18 1,4-Dehydrogenation.

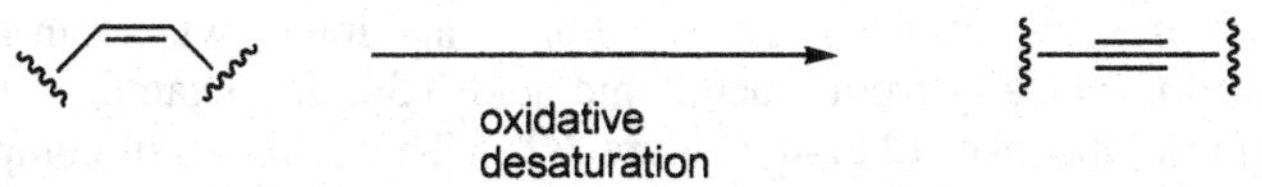

Figure 3.19 Formation of acetylenes.

atoms located at positions 8 and 11 in linoleic acid (i.e. they are in a 1,4-relationship) are removed giving the diene fragment between positions 8 and 11 in calendic acid. Note the *trans* configuration of two double bonds in this diene.

We finally mention in this section acetylenic fatty acid derivatives such as crepenynic acid **3.3**. It is generally believed that the acetylene group is introduced by further oxidative desaturation of an existing double bond (Figure 3.19). Many plants, particularly Umbeliferae and Compositae, are sources of polyacetylene natural products e.g. dehydromatricaria acid **3.36** and cicutoxin **3.37** (Figure 3.20). Natural products containing the thiophene or furan ring systems are often isolated with polyacetylenes, and these heterocycles are presumably formed by addition of hydrogen sulphide (or an equivalent species) and water respectively across two conjugated acetylene units (Figure 3.21).

3.36

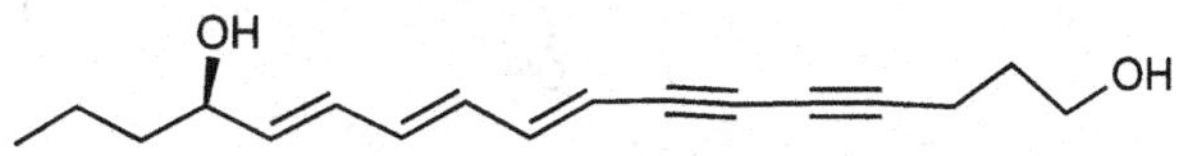

3.37

Figure 3.20 Polyacetylene natural products.

thiophenes (X = S)
furans (X = O)

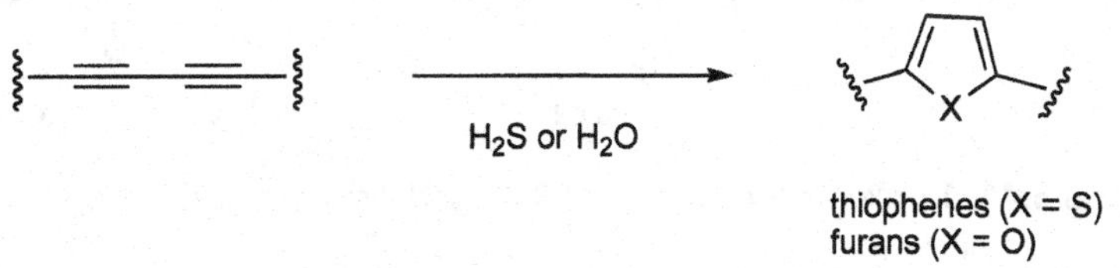

Figure 3.21 Formation of thiophene and furan natural products.

3.7 Oxygenated Fatty Acids

Oxygenated fatty acids (which are often referred to as *oxylipins*) are found widely in both plants and animals. In mammals, oxylipins are derived principally from arachidonic acid **3.34**, and examples include leukotriene A$_4$ **3.38**, prostaglandin PGF$_{2\alpha}$ **3.39** and thromboxane A2 **3.40**. (Figure 3.22). These classes of compounds exert a diverse range of pharmacological effects in mammals including regulation of blood pressure, control of blood platelet aggregation, allergic responses and inflammation processes. Linolenic acid **3.41** is frequently the source of oxylipins in plants and is the precursor of 12-oxophytodienoic acid **3.42**, which is in turn the precursor of the widely distributed hormone, jasmonic acid **3.43** (Figure 3.23). In plants oxylipins have a multitude of functions, for example they are involved in wound healing, defence mechanisms against pathogens and are components of cuticle.

The biosynthetic pathway to oxylipins often involves the addition of oxygen to an unsaturated fatty acid (R—H) to give a hydroperoxide i.e. R—OOH. This reaction is catalysed by enzymes called *lipoxygenases* – these enzymes are a type of dioxygenase because they introduce both oxygen atoms from oxygen into the product. The hydroperoxides can then undergo a series of reactions to give the final oxylipin.

The biosynthesis of leukotriene A$_4$ **3.38** is shown in Figure 3.24. Arachidonic acid **3.34** and oxygen are converted into a hydroperoxide **3.44** by a 5-lipoxygenase. Note the use of *fish-hooks* in the reaction mechanism because unpaired electrons i.e. radicals are involved. Molecular oxygen is a diradical (it has two unpaired electrons in its electronic ground state).

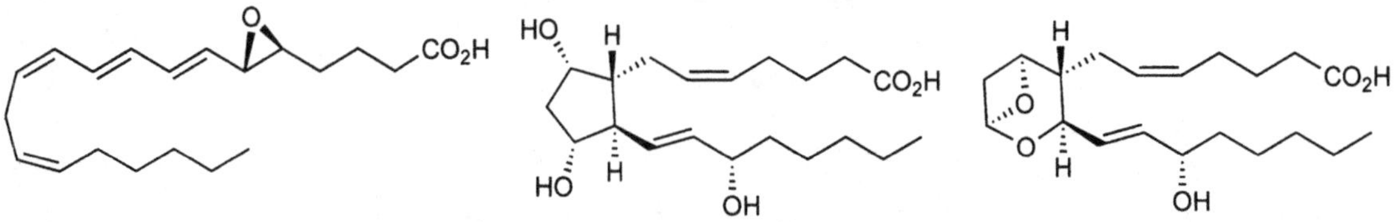

Figure 3.22 Leukotriene A$_4$ **3.38**, prostaglandin PGF$_{2\alpha}$ **3.39** and thromboxane A2 **3.40**.

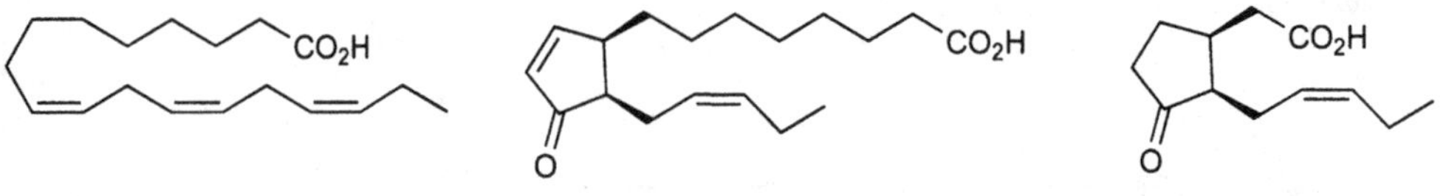

Figure 3.23 Linolenic acid **3.41**, 12-oxophytodienoic acid **3.42** and jasmonic acid **3.43**.

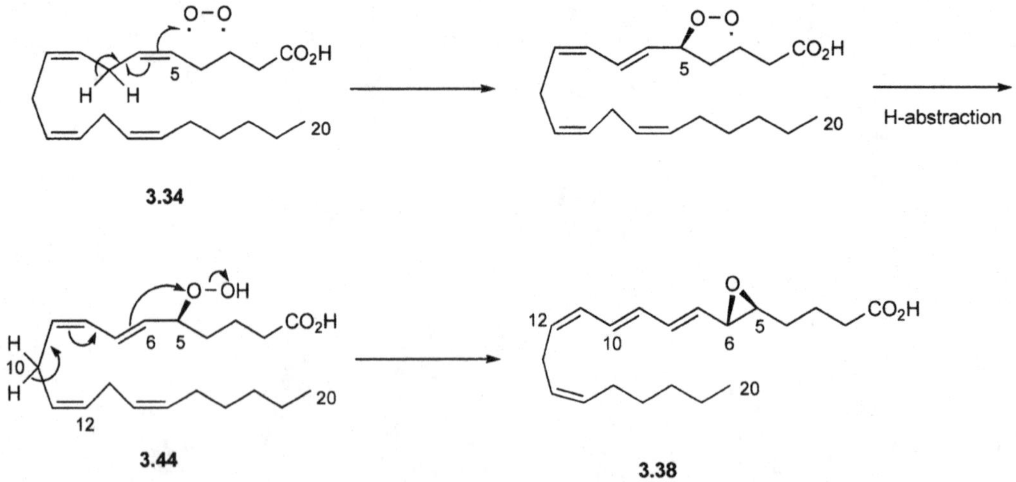

Figure 3.24 Biosynthesis of leukotriene A$_4$ **3.38**.

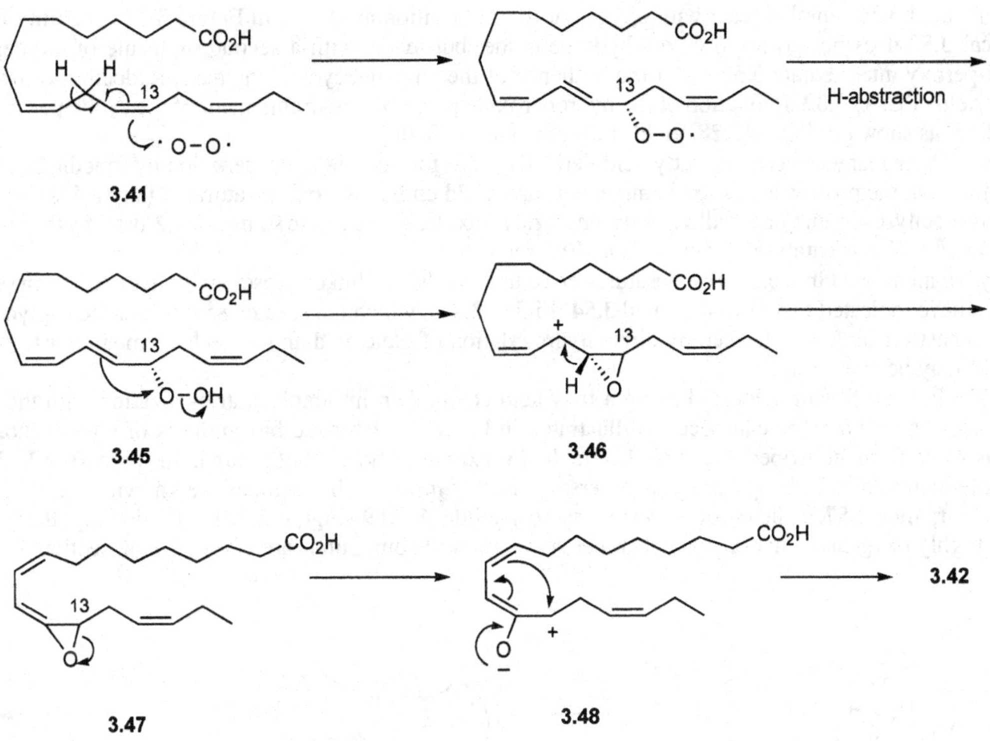

Figure 3.25 Biosynthesis of 12-oxophytodienoic acid **3.42**.

The biosynthesis of 12-oxophytodienoic acid **3.42** is shown in Figure 3.25. Linolenic acid **3.41** and oxygen give a hydroperoxide **3.45** in the presence of a 13-lipoxygenase and this reaction is mechanistically similar to the hydroperoxidation of arachidonic acid **3.34**. This hydroperoxide then gives the intermediate carbocation **3.46** (note the allylic stabilization) which in turn furnishes the epoxide **3.47**. In this respect the mechanism is different from the one shown in Figure 3.24. Ring-opening of the epoxide then yields a dipolar intermediate **3.48** (note that the positive charge can be delocalized) which subsequently cyclizes giving 12-oxophytodienoic acid **3.42**. The hydroperoxide is also the precursor to several other natural products. For example, a carbon to oxygen migration can occur eventually yielding a mixture of 12-oxo-(9Z)-dodecenoic acid **3.50** and (3Z)-hexenal **3.51** via the intermediate **3.49** (Figure 3.26). Many C₆-aldehydes and alcohols are produced by plants using related pathways. These *short-chain* compounds are responsible for the *green odour* of many plants.

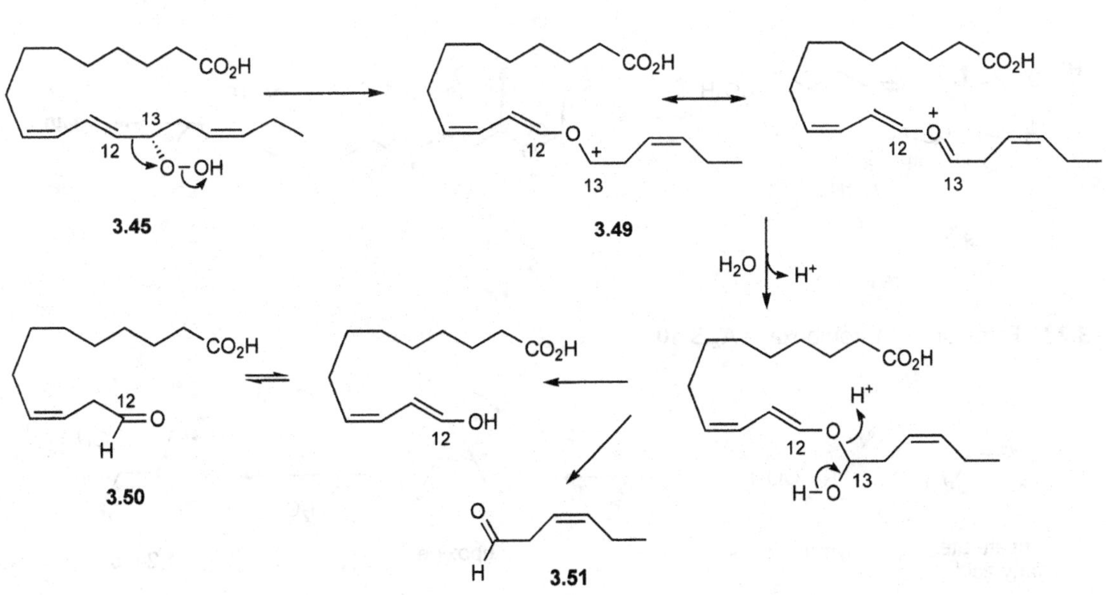

Figure 3.26 Biosynthesis of the acid **3.50** and the aldehyde **3.51**.

Arachidonic acid **3.34** can also react with oxygen at the 11-position as shown in Figure 3.27. The initially formed peroxy radical **3.52** does not go on to give a hydroperoxide, but reacts with a second molecule of oxygen to yield the cyclic di-peroxy intermediate **3.53** under the influence of the enzyme, cyclooxygenase. Reduction of both peroxy groups then yields PGF$_{2\alpha}$ **3.39**. Reduction of the hydroperoxide group and rearrangement of the cyclic peroxy group in compound **3.53**, as shown in Figure 3.28, yields thromboxane A$_2$ **3.40**.

We have seen how many oxygenated fatty acid derivatives are formed via hydroperoxide intermediates (R—OOH). Hydroperoxides can themselves act as oxidizing agents and yield epoxides from unsaturated fatty acids (Figure 3.29). This reaction is catalysed by enzymes called *peroxygenases*. Epoxides are often the source of 1,2-diols by the nucleophilic addition of water, which is catalysed by *epoxide hydrolases*.

Previously we mentioned that oxidative desaturation of fatty acids was linked closely with hydroxylation (see Figure 3.14). Studies have indicated that ricinoleic acid **3.54** (Figure 3.30), which constitutes 85–95% of the triglyceride fatty acids found in castor bean oil, is formed by direct hydroxylation of oleic acid and a mechanism similar to that shown in Figure 3.14 may be operating.

When an alcohol substituent is located along a fatty acid chain, then intramolecular cyclization with the fatty acid thioester group to give a *lactone* can occur as illustrated in Figure 3.31 for the biosynthesis of vittatalactone **3.56**, a pheromone isolated from the striped cucumber beetle. In this example, the alcohol group in the precursor **3.55** has been produced by reduction of a carbonyl group. A diverse variety of macrocyclic lactones are known, of which examples include microcarpalide **3.57**, colletodiol **3.58** and macrosphelide A **3.59** (Figure 3.32). Additionally, there are many examples of highly oxygenated macrocyclic lactones some of these compounds are considered in section 4.

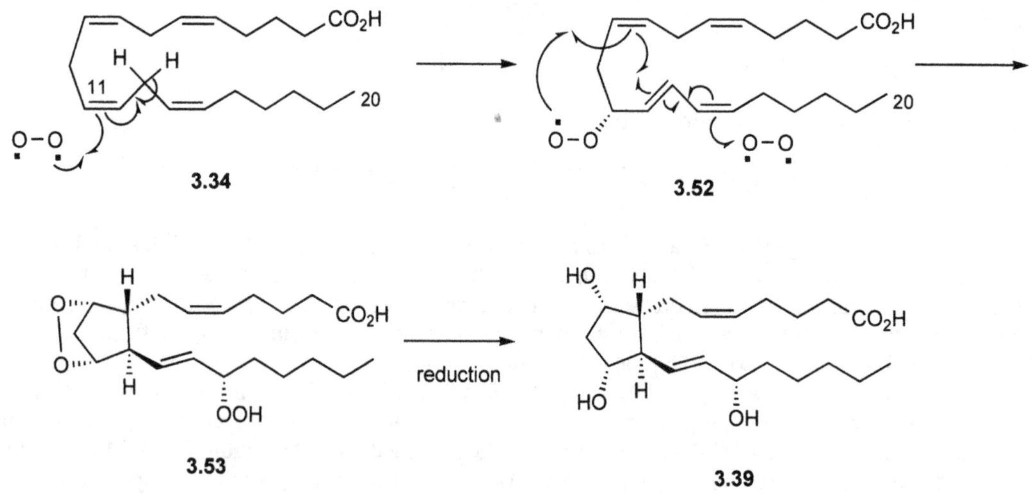

Figure 3.27 Formation of PGF$_{2\alpha}$ **3.39**.

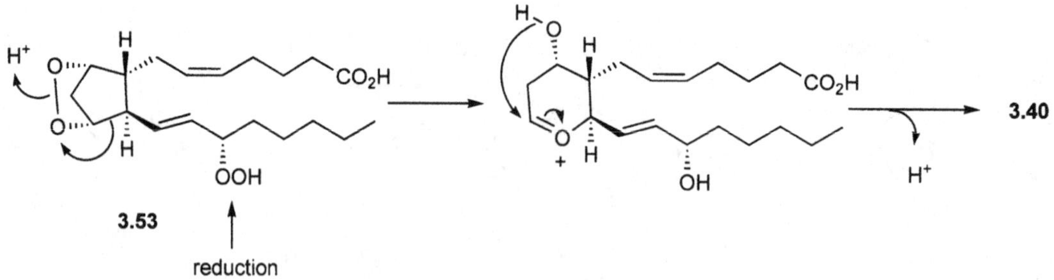

Figure 3.28 Formation of thromboxane A$_2$ **3.40**.

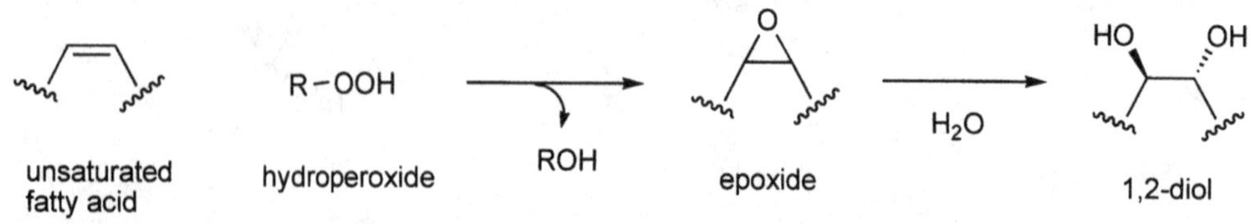

Figure 3.29 Formation of epoxides and 1,2-diols.

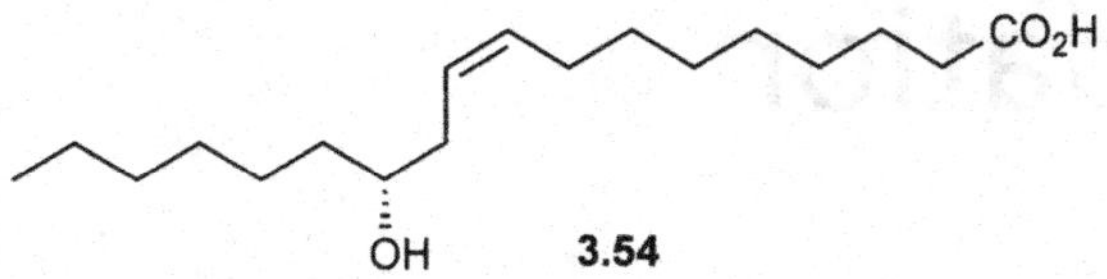

Figure 3.30 Ricinoleic acid **3.54**.

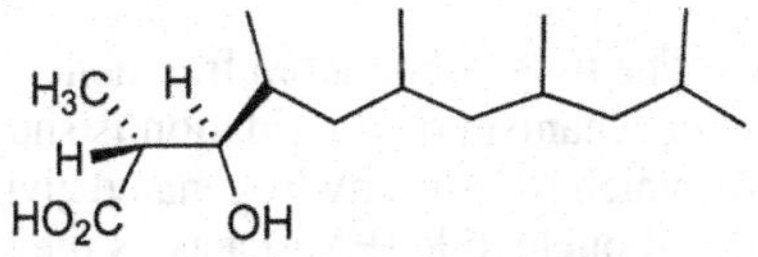

Figure 3.31 Vittatalactone **3.56** and its precursor **3.55**.

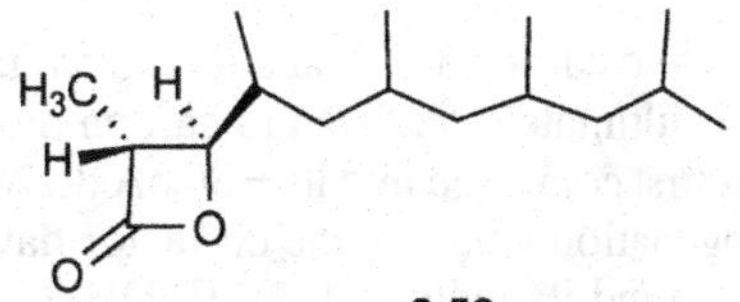

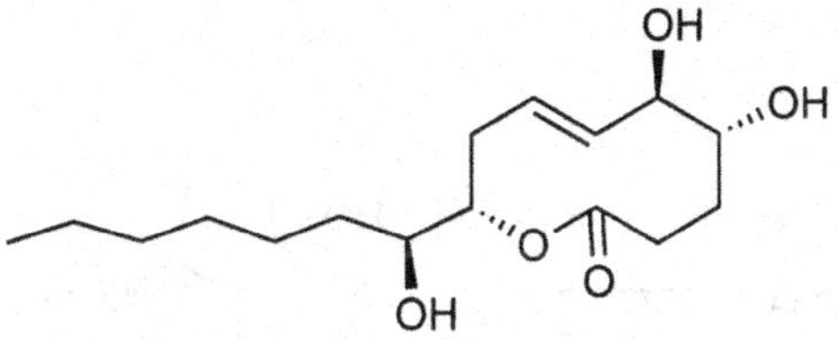

Figure 3.32 Macrocyclic lactones.

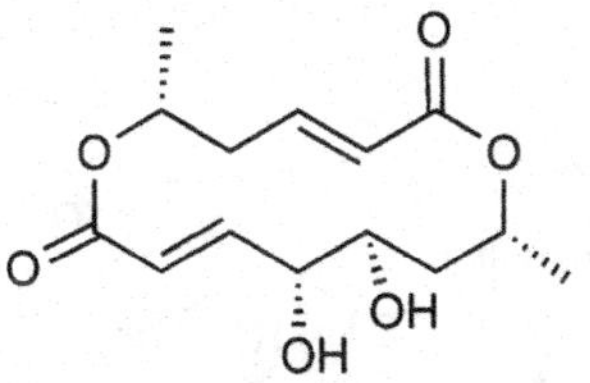

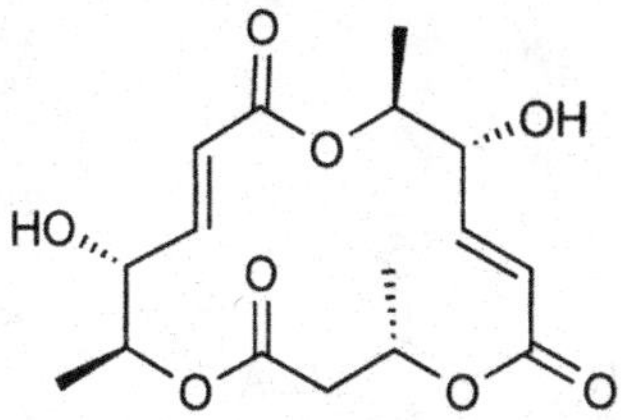

Problem

Problem 3.8 Identify the hydroxy-acid components of colletodiol **3.58** and macrosphelide A **3.59**.

In addition to intramolecular cyclization yielding lactones, hydroxyl derivatives of fatty acids can also participate in intermolecular condensation reactions giving polyhydroxyalkanoates (PHAs). These compounds are polyesters that are produced as an energy reserve by a variety of bacteria. They have attracted considerable interest as sources of biodegradable polymers because numerous bacteria are capable of degrading extra-cellular PHAs. Poly(3-hydroxybutyrate) **3.60** is an example of a PHA and is biosynthesized by polymerization of compound **3.11** (Figure 3.33). There has also been recent interest in using genetic engineering as a means of introducing functional groups at one end of the PHA which can interact with specific receptors or ligands to produce novel block co-polymers.

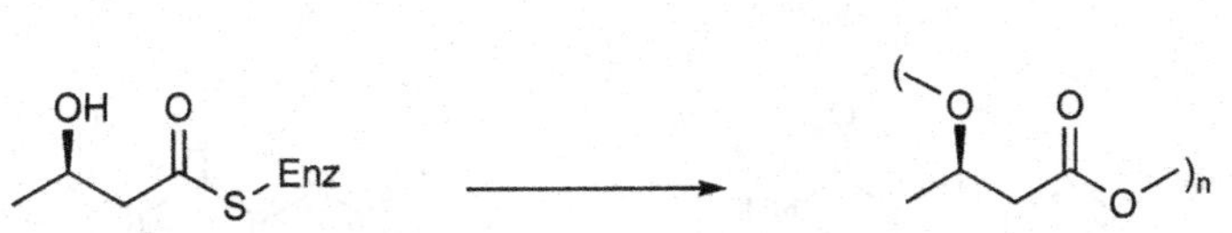

Figure 3.33 Biosynthesis of poly(3-hydroxybutyrate) **3.60**.

31

3.8 β-Oxidation

β-Oxidation removes two carbon atoms at a time from fatty acids. The two-carbon atom fragment is removed as acetyl coenzyme A which is ultimately oxidized to carbon dioxide. The mechanism of β-oxidation is shown in Figure 3.34. Thus, the fatty acid is first converted into its —SCoA derivative **3.61** which is then dehydrogenated giving the *trans* alkene **3.62**. In the dehydrogenation reaction, the co-factor flavin adenine dinucleotide (FAD) acts as the hydrogen acceptor. The structure of FAD (and its reduced form FADH$_2$) is shown in Figure 3.35. Alkene **3.62** is then stereospecifically hydrated (this can be viewed as a conjugate addition of water to the alkene) giving the (S)-alcohol **3.63**. The alcohol **3.63** is then oxidised by NAD$^+$ giving the β-keto thioester derivative **3.64**. In this oxidation, hydride is transferred from the alcohol to NAD$^+$ giving NADH: this is the reverse of the reaction in which NADPH acts as a hydride donor.

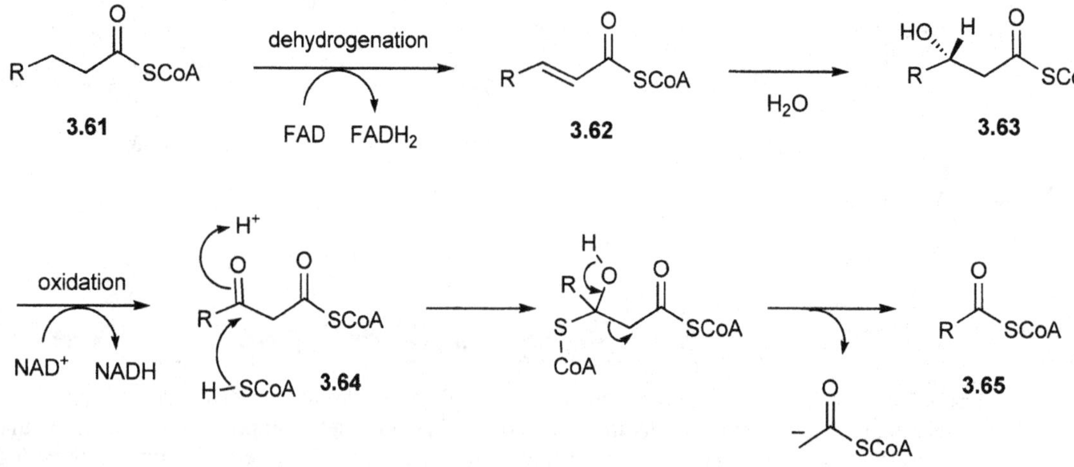

Figure 3.34 β-Oxidation.

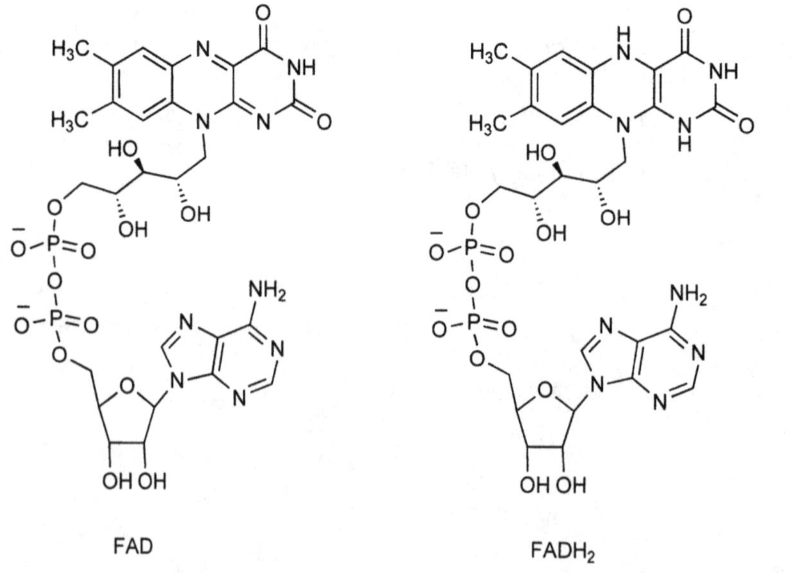

Figure 3.35 Flavin adenine dinucleotide (FAD) and its dihydro derivative (FADH$_2$).

Note that in β-oxidation NAD$^+$/NADH is used rather than NADP$^+$/NADPH. Finally, nucleophilic attack of HSCoA at the carbonyl group in compound **3.64** and expulsion of the anion of acetyl coenzyme A gives the shortened fatty acid **3.65**.

There are many examples in which natural products have been biosynthesized by pathways that involve β-oxidation. Some representative examples from this section include the biosynthesis of jasmonic acid **3.43** from 12-oxophytodienoic acid **3.42** and the formation of dehydromatricaria acid **3.36** from oleic acid **3.2** (note the isomerization of the double bond at some stage).

Cinnamic acid **3.66** and its derivatives are produced as part of the shikimic acid pathway (section 5). β-Oxidation of cinnamic acid derivatives yields benzoic acid derivatives **3.67** (Figure 3.36).

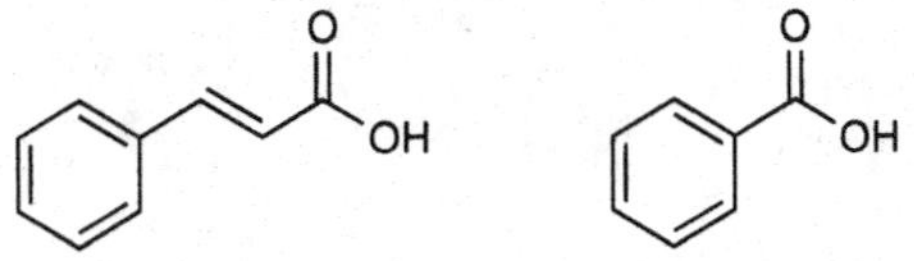

3.66   **3.67**

Figure 3.36   Cinnamic acid **3.66** and benzoic acid **3.67**.

Summary

(a) Even-numbered straight-chain saturated fatty acids are biosynthesized by elongation of an acetyl starter unit with malonyl units.

(b) Odd-numbered straight-chain fatty saturated acids are biosynthesized by elongation of a propionyl starter unit with malonyl units.

(c) Branched starter units or substituted malonyl units give branched saturated fatty acids.

(d) Mono-unsaturated fatty acids are produced from saturated precursors by oxidative desaturation, generally with production of a *cis* double bond between carbon atoms 9 and 10.

(e) Polyunsaturated fatty acids and acetylenic fatty acids are produced by further oxidative desaturation of unsaturated precursors.

(f) The addition of oxygen to unsaturated fatty acids gives hydroperoxides which are the precursors of many oxygenated fatty acid derivatives.

(g) The coenzyme A derivatives (RCO—CoA) of fatty acids can react with a range of nucleophiles to produce esters, lactones, amides, aldehydes and alcohols.

Biosynthesis of Polyketides

Objectives

After studying this section you will be able to:

(a) Recognize the structure of natural products derived from the polyketide pathway.
(b) Describe the primary process in which polyketides are biosynthesized.
(c) Recognize the various cyclization pathways encountered in the formation of polyketides.
(d) Describe how secondary processes give further structural diversification.
(e) Outline how isotopes can be used to establish biosynthetic pathways.

4.1 What are Polyketides?

In section 2, Figure 2.11 we saw how acetyl coenzyme A and malonyl coenzyme A could be combined giving a thioester intermediate **4.1** (Figure 4.1) where the 'Enz' fragment is the acyl carrier protein. In structure **4.1**, the atoms labelled *a* are derived from acetyl coenzyme A and the atoms labelled *m* are derived from malonyl coenzyme A. In fatty acid biosynthesis (section 3), the ketone group of thioester **4.1** is reduced giving compound **4.2** which is the precursor of numerous fatty acids.

Instead of reduction of thioester **4.1** to compound **4.2**, the ketone group in thioester **4.1** can be retained. If thioester **4.1** reacts with a molecule of malonyl coenzyme A (Figure 4.2) the thioester **4.3** is produced. This reaction follows a similar mechanism to that which has already been discussed in section 3 for fatty acid biosynthesis (see Figure 3.6 section 3 for example). The thioester **4.3** now contains two ketone carbonyl groups and one thioester carbonyl group. Further addition of a malonyl coenzyme A unit to thioester **4.3** then gives a product **4.4**. Structures such as **4.3** and **4.4**, which possess a chain of alternating ketone and methylene groups ($>CH_2$) are called *polyketides*. Notice also that each methylene group in a polyketide is flanked by two carbonyl groups ($>C=O$). The structural unit $-CO-CH_2-CO-$ is called a *1,3-dicarbonyl* or *β-dicarbonyl* and, as we shall see in section 4.3, the 1,3-dicarbonyl fragments of polyketides play an important role in their chemistry. In structure **4.4**, the atoms labelled *a* have been derived from acetyl coenzyme A and the atoms labelled *m1*, *m2* and *m3* have been derived from malonyl coenzyme A units.

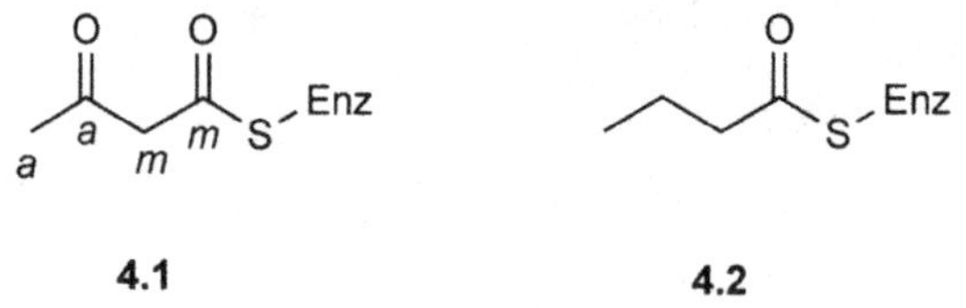

Figure 4.1 Thioesters.

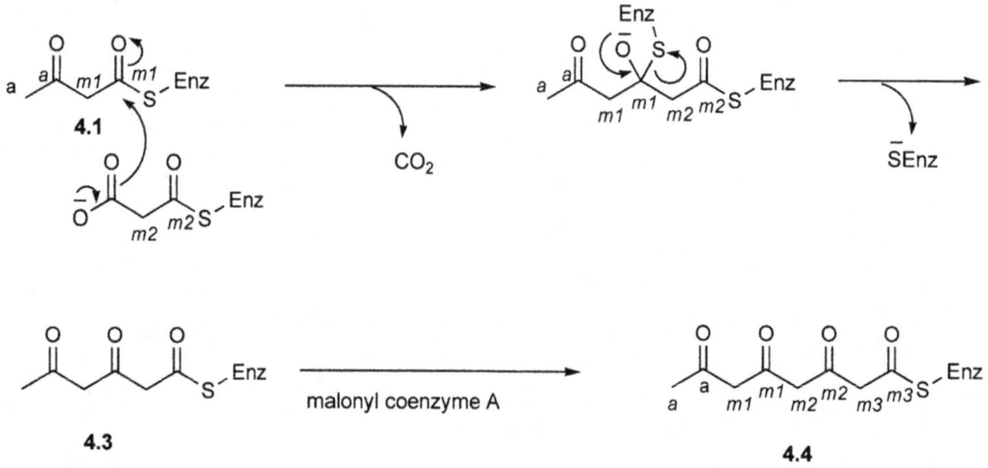

Figure 4.2 Polyketides.

Problem

Problem 4.1 Draw the structure of the polyketide you would expect to obtain from propionyl coenzyme A and three molecules of methyl malonyl coenzyme A.

4.2 The Chemistry of 1,3-Dicarbonyls: Keto–Enol Tautomerism

Ethyl acetoacetate **4.5** (Figure 4.3) is a typical 1,3-dicarbonyl compound and is an ester analogue of the thioester **4.1**. In fact, as we shall see later, the reactions of esters and their thioester analogues are closely related. Although structure **4.5** is commonly drawn as a 1,3-dicarbonyl compound, its structure can also be represented by the enol tautomer **4.6** (the term enol indicates both alkene and alcohol functional groups). Simple ketones such as acetone exist mainly as their keto tautomers **4.8** rather than their corresponding enol tautomers **4.9**. In 1,3-dicarbonyl compounds such as compound **4.5**, the concentration of the enol form **4.6** is generally much greater than in simple ketones such as acetone. This is because the oxygen lone pair of the hydroxy group can be delocalized through to the ester carbonyl group giving structure **4.7**. It is this delocalization of electrons that accounts for the presence of the enol form of 1,3-dicarbonyl compounds. Intramolecular hydrogen-bonding between the hydroxy group and the ester group, as shown by the dashed line in formula **4.6**, is a factor that also favours the enol tautomer.

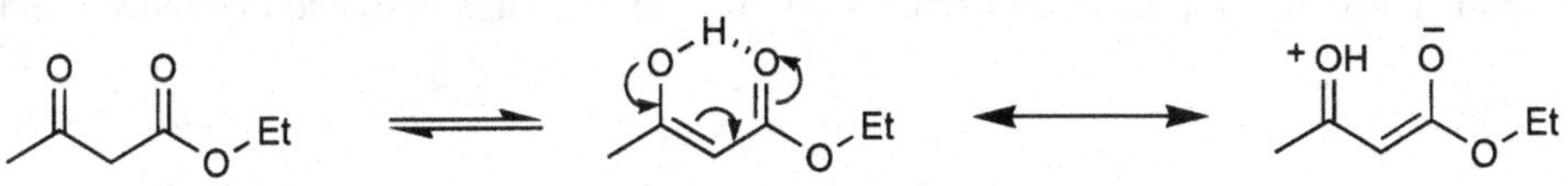

4.5	**4.6**	**4.7**
keto tautomer	enol tautomer	

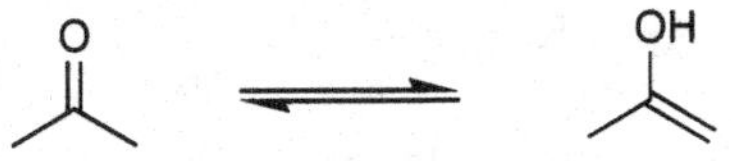

4.8 **4.9**

Figure 4.3 Keto and enol tautomers.

4.3 The Chemistry of 1,3-Dicarbonyls: Condensation Reactions

One particularly well known organic reaction is the condensation reaction between 1,3-dicarbonyl compounds (which probably react as their enol tautomers) and aldehydes (the *Knoevenagel* reaction). In this reaction, a new carbon–carbon double bond is formed and water is produced. An example of this type of reaction involving the enol form **4.6** of the 1,3-dicarbonyl compound **4.5**, together with a mechanism, is shown in Figure 4.4. The overall reaction between a 1,3-dicarbonyl compound and an aldehyde can therefore be represented as shown in Figure 4.5. Note that both the protons of the methylene group are eventually lost as water. The Knoevenagel reaction is usually catalysed by bases or acids but these have been omitted for simplicity. Although we have illustrated the Knoevenagel reaction using aldehydes, in principle ketones can also participate in this reaction. Note that in many texts this reaction is often called an *aldol* reaction; both the aldol and Knoevenagel reactions are closely related.

A second type of condensation reaction of 1,3-dicarbonyls which is relevant to polyketide chemistry is their reaction with esters. This reaction is related to the well known organic reaction called the *Claisen* condensation. The mechanism of this condensation is illustrated in Figure 4.6 for the reaction of the 1,3-dicarbonyl compound **4.5** and an ester. You

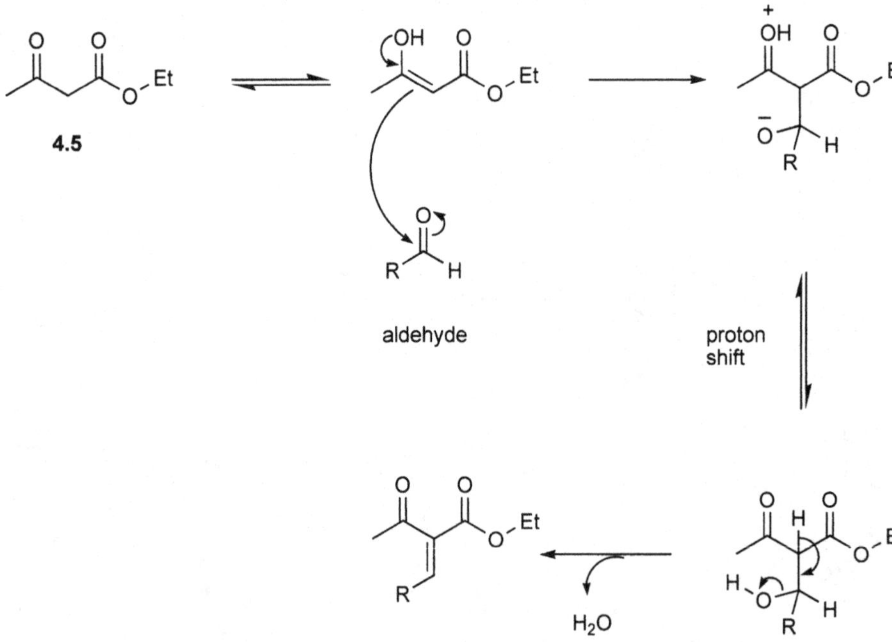

Figure 4.4   The Knoevenagel reaction.

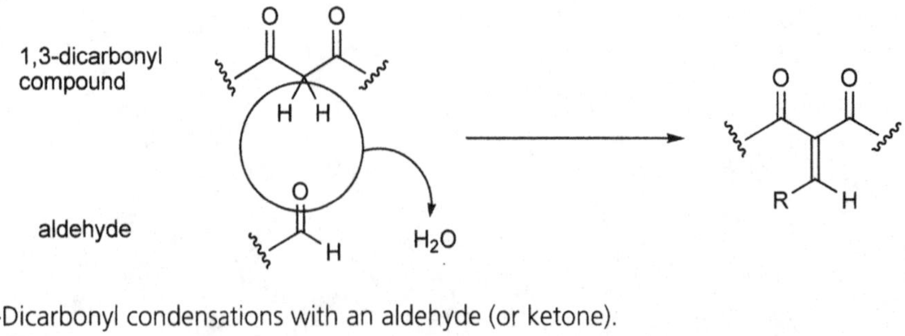

Figure 4.5   1,3-Dicarbonyl condensations with an aldehyde (or ketone).

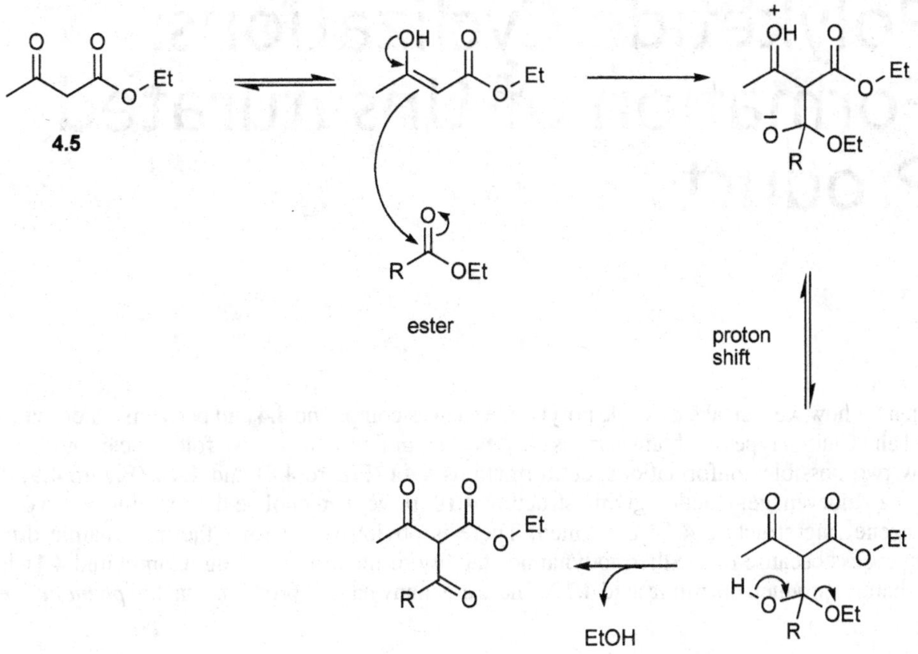

Figure 4.6 The Claisen condensation.

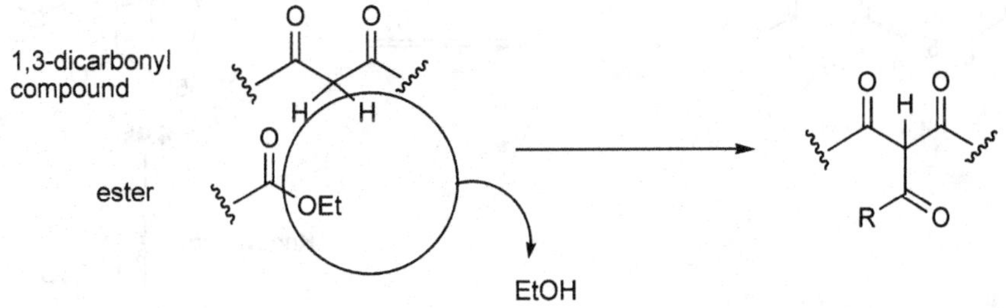

Figure 4.7 1,3-Dicarbonyl condensations with esters.

should compare this reaction with that shown in Figure 4.4. Note that in Claisen-type reactions an *alcohol* (in this case ethanol) is lost whereas in Knoevenagel-type reactions *water* is lost. The overall reaction between a 1,3-dicarbonyl compound and an ester can therefore be represented as shown in Figure 4.7 and you should compare this reaction with the Knoevenagel reaction shown in Figure 4.5. The Claisen reaction generally requires the presence of a base but this has been omitted from Figure 4.6 for simplicity.

In summary, there are two important types of condensation reactions that are relevant to polyketide chemistry; those in which water is lost (as previously illustrated for the reaction of 1,3-dicarbonyls and aldehydes) and those in which an alcohol is lost (as previously illustrated for the reaction of 1,3-dicarbonyls and esters).

In the reaction illustrated in Figure 4.7, an ester has been condensed with a 1,3-dicarbonyl compound. In principle, *a thioester could replace the ester* in Figure 4.7 giving the same product but with loss of thiol (RSH) rather than an alcohol. This is important in biosynthesis as Nature often works with thioesters.

4.4 Polyketide Cyclizations: Formation of Unsaturated Products

We will now consider how we can take a simple polyketide such as compound **4.4** and perform condensation reactions of the Knoevenagel and Claisen types to obtain various classes of natural products. If we fold the carbon chain of compound **4.4** we can draw two possible conformations, conformations **4.4a** (Figure 4.8) and **4.4b** (Figure 4.9). Conformation **4.4a** can undergo a Knoevenagel reaction giving structure **4.10**. If we then enolize the remaining two carbonyl groups, an aromatic benzene ring structure **4.11** is formed. There is obviously a strong thermodynamic driving force for this enolization process because of the thermodynamic stability of the aromatic ring. Compound **4.11** is the thioester derivative of the natural product, orsellinic acid **4.12**. There are many natural products that are *phenolic* (i.e. they contain

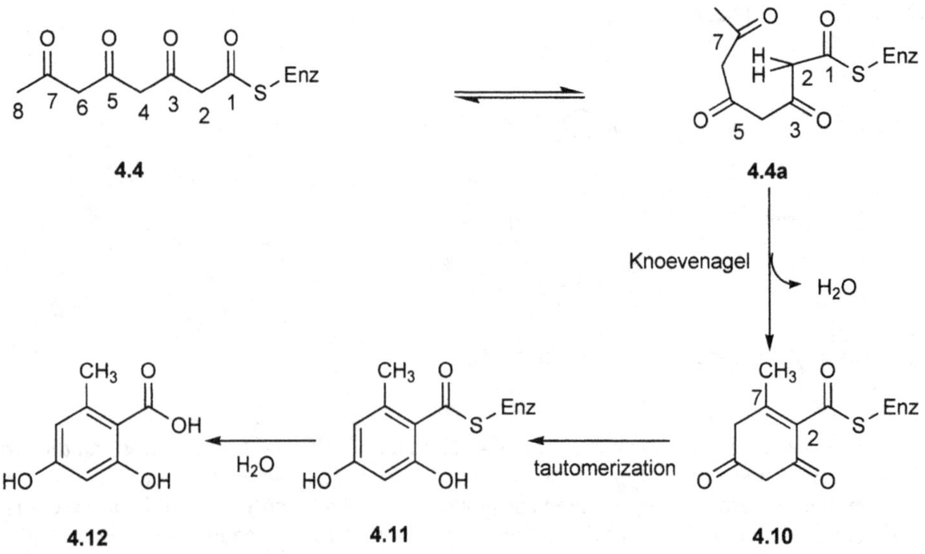

Figure 4.8   Biosynthesis of orsellinic acid **4.12**.

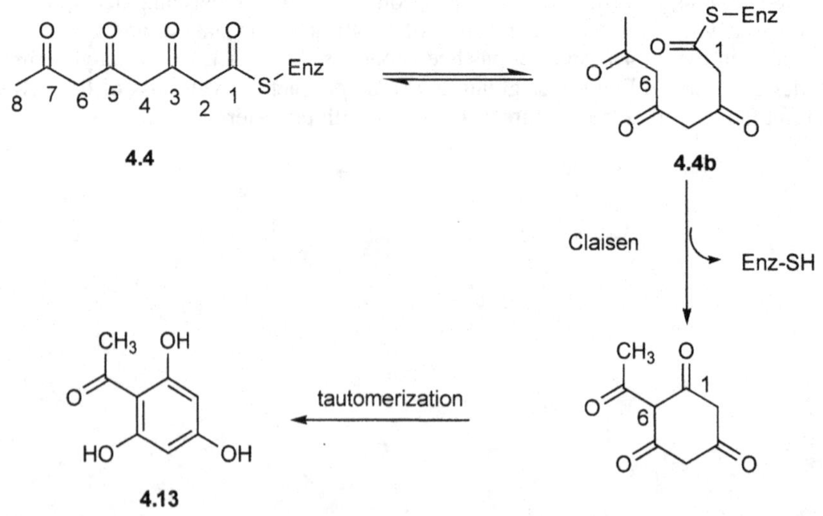

Figure 4.9   Biosynthesis of xanthoxylin **4.13**.

a benzene ring substituted with one or more hydroxyl groups), and many are formed by biosynthetic processes related to the one shown in Figure 4.8. Figure 4.9 depicts the cyclization of conformation **4.4b** via an initial Claisen condensation giving xanthoxylin **4.13**. Not all phenolic materials are derived from polyketides, some products are biosynthesized from shikimic acid (see section 5).

Problem

Problem 4.2 In Figures 4.8 and 4.9 cyclization has occurred with formation of carbon–carbon bonds. Carbon–oxygen bonds may also result from polyketide cyclizations. Deduce how the polyketide **4.3** might give the 2-pyrone derivative depicted below?

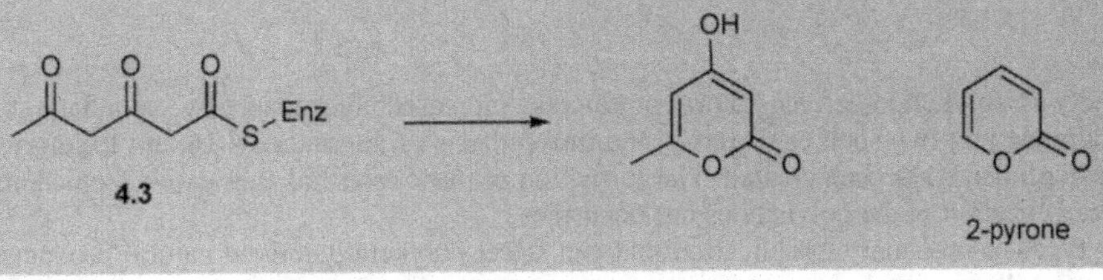

The biosyntheses of orsellinic acid **4.12** and xanthoxylin **4.13** described above illustrate two possible modes of cyclizing the polyketide **4.4**. With longer polyketides, there are potentially more cyclization pathways. Figure 4.10 illustrates a possible biosynthetic pathway to the natural product alternariol **4.16** from the polyketide **4.14**. This biosynthetic pathway involves two Knoevenagel-type condensations and in the last step shown in Figure 4.10, an ester linkage is formed between the phenolic hydroxyl group and the thioester group in compound **4.15**. Cyclic esters such as alternariol are called *lactones*.

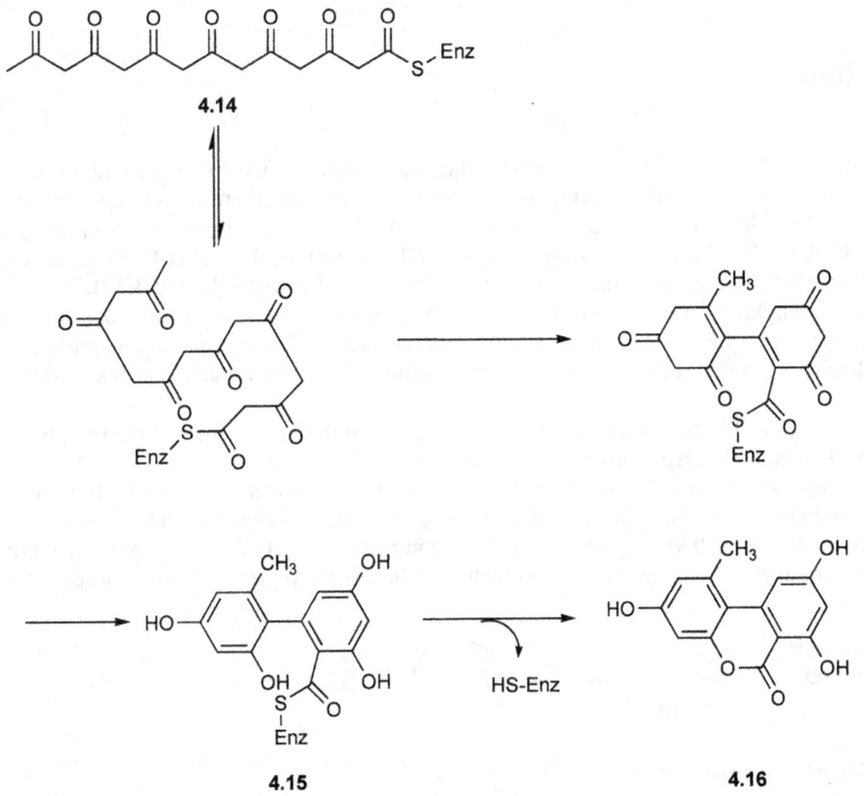

Figure 4.10 Biosynthesis of alternariol **4.16**.

Problem

Problem 4.3 Lecanoric acid, empirical formula $C_{16}H_{14}O_7$, is an ester formed from the condensation of two molecules of orsellinic acid **4.12**. Draw the two possible structures for lecanoric acid.

4.5 Secondary Structural Modifications During Polyketide Cyclizations

The structural variety of polyketide-derived natural products is increased enormously by secondary structural modifications. We have already met two such examples in the biosynthesis of alternariol **4.16** and lecanoric acid (Problem 4.3), in which an ester linkage has been created. The formation of these ester linkages can be considered as secondary modifications after cyclization of the polyketides has occurred.

There are many types of secondary modification that can affect polyketide-derived natural products. Five common modifications that we will consider here are:

- Alkylation
- Reduction
- Oxidation
- Decarboxylation
- Modifications to the carbon skeleton

4.5.1 Alkylation

Methylation

Two general types of methylation reaction can occur in biological systems which are relevant to this section. These are (i) methylation of phenolic hydroxyl groups creating a *carbon–oxygen* bond and therefore giving the corresponding methyl ethers i.e. Ar—OH → Ar—OMe (Ar = aromatic group) and (ii) methylation at a carbon atom creating a *carbon–carbon* bond, i.e. C—H → C—CH$_3$. The biological methylating agent that is responsible for both these reactions is *S*-adenosyl methionine, a molecule we have already encountered in section 3 (see Figures 3.11 and 3.12).

The mechanism of methylation at oxygen (*O*-methylation) is shown in Figure 4.11 and closely follows that of typical organic methylations using methylating agents, such as methyl iodide in the presence of an appropriate base. Note that phenolic hydroxyl groups are very acidic and are therefore deprotonated by relatively weak bases such as potassium carbonate.

Figure 4.12 shows a plausible biosynthetic pathway to eugenone **4.18** from the polyketide precursor **4.17**. In this biosynthesis, three *O*-methylation reactions have occurred.

1,3-Dicarbonyl compounds are readily methylated at the methylene carbon atom (*C*-methylation). The mechanism of this reaction is shown in Figure 4.13 and involves the corresponding enol tautomer of the 1,3-dicarbonyl compound and *S*-adenosyl methionine as the methylating agent. This reaction is also closely related to organic methylation reactions of 1,3-dicarbonyls using methylating agents such as methyl iodide in the presence of an appropriate base.

Chemical methylation

Biological methylation

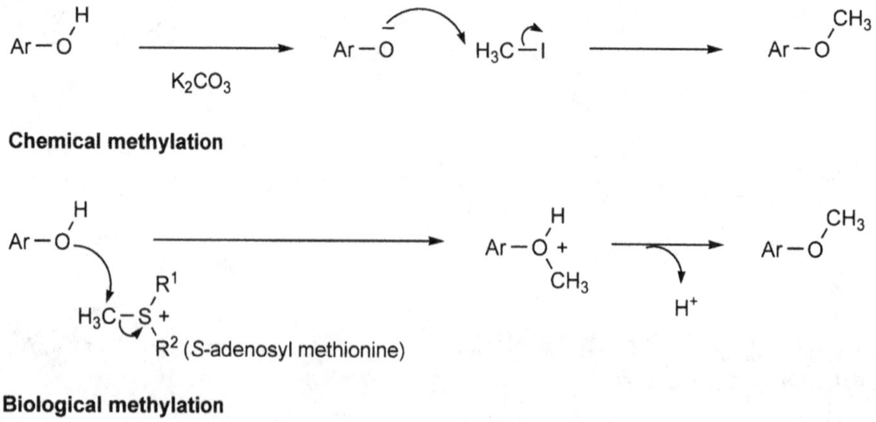

Figure 4.11 Methylations at oxygen.

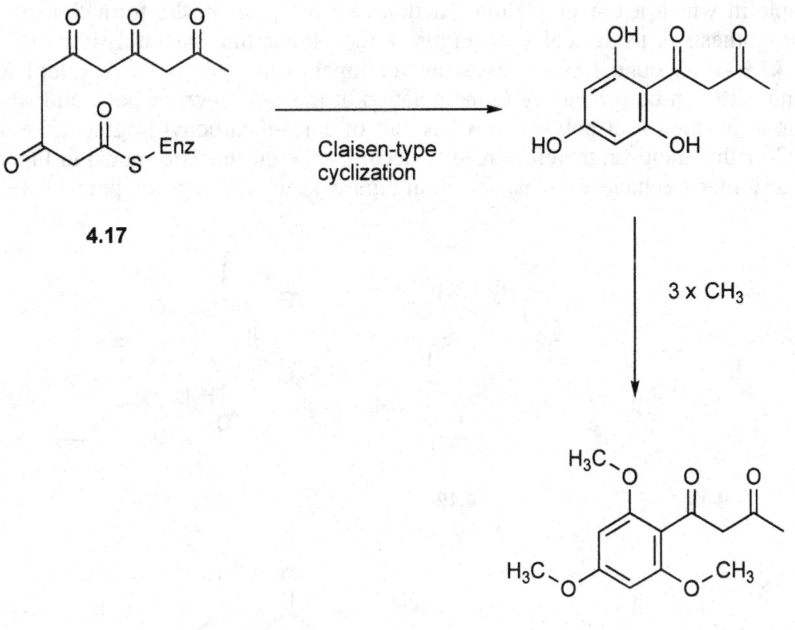

Figure 4.12 Biosynthesis of eugenone **4.18**.

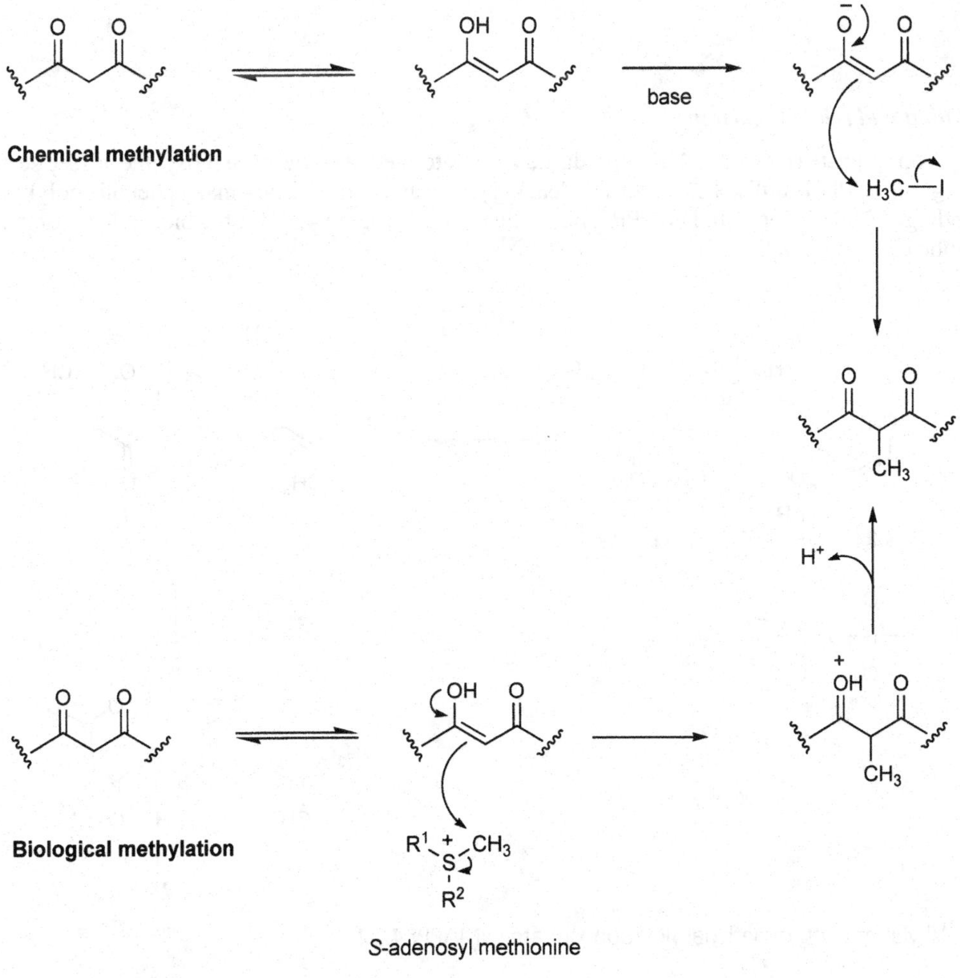

Figure 4.13 Methylations at carbon.

One relevant example in which a *C*-methylation reaction is involved is in the formation of compound **4.19**, an intermediate in the biosynthesis of usnic acid **4.20** (Figure 4.14). Notice the structural similarity between compound **4.19** and xanthoxylin **4.13** – compound **4.19** possesses an additional methyl group. If you refer back to Figure 4.9 you will see that compound **4.19** can be formed by *C*-methylation at the 4-position of polyketide **4.4** giving polyketide **4.21**. The carbon atom at position 4 of polyketide **4.4** is part of a 1,3-dicarbonyl fragment, i.e. it is flanked by two carbonyl groups and *C*-methylation can therefore readily occur by the mechanism shown in Figure 4.13. Cyclization of polyketide **4.21** by a similar mechanism to that shown in Figure 4.9 would give compound **4.19**.

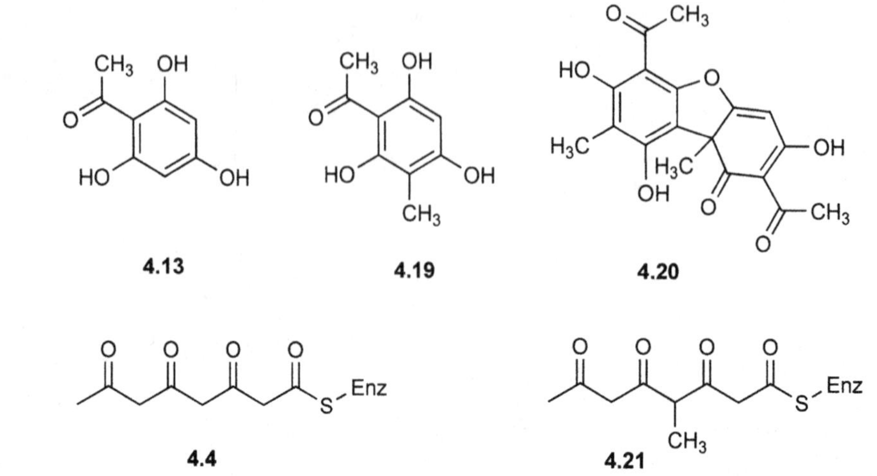

Figure 4.14 Methylated products.

Dimethylallylation and related reactions

Dimethylallylpyrophosphate (DMAPP) **4.22** is produced as an intermediate in terpene biosynthesis (see section 5). The pyrophosphate group (OPP) is utilised in Nature as a leaving group and consequently many phenolic polyketide-derived compounds undergo *C*-alkylation with DMAPP. This is illustrated in Figure 4.15 for the biosynthesis of peucenin **4.24** from the phenolic compound **4.23**.

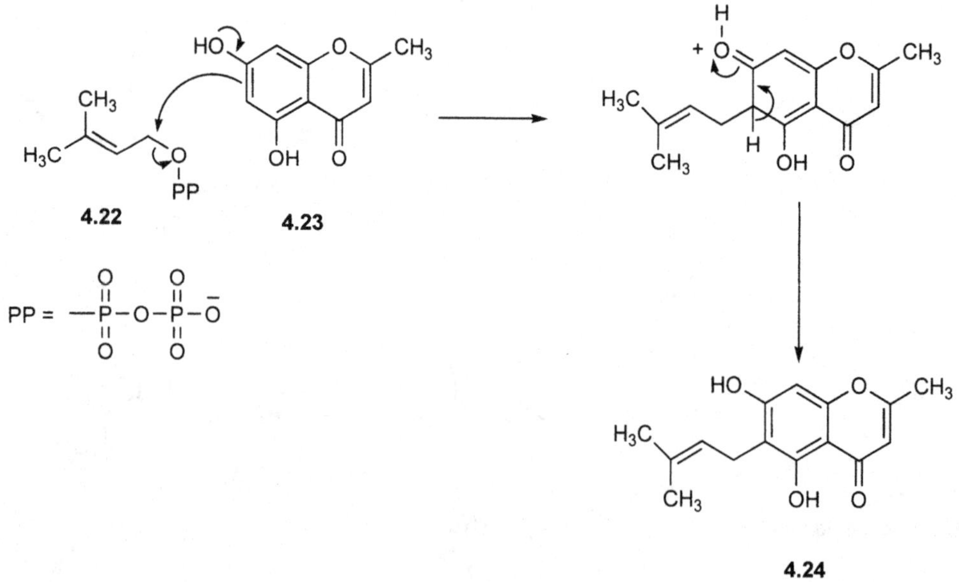

Figure 4.15 Alkylation with dimethylallylpyrophosphate (DMAPP) **4.22**.

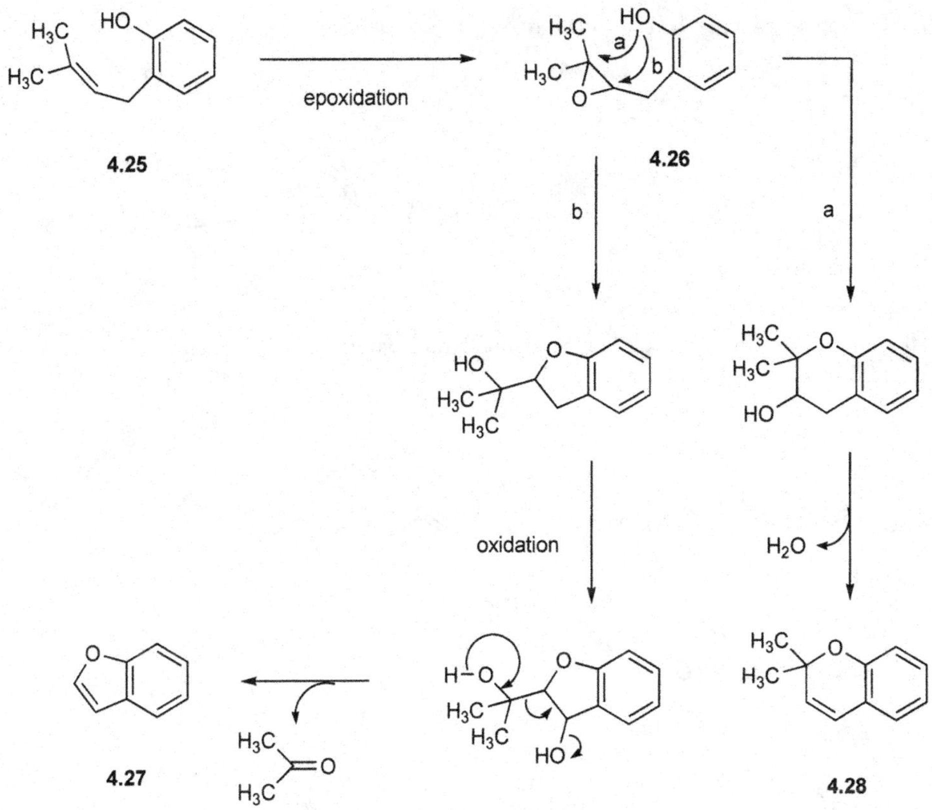

Figure 4.16 Formation of benzofuran **4.27** and dihydrochromene **4.28** rings.

Natural products such as peucenin can be considered as products of mixed biosynthetic origin, i.e. part of the molecule originates from the polyketide pathway and part of the compound is derived from the terpene pathway. Many natural products are derived from precursors of two or more biosynthetic origins.

The dimethylallyl group in dimethylallyated phenols, as represented by the structure **4.25**, can undergo a range of further transformations leading to the formation of oxygen-containing rings (Figure 4.16). Thus, epoxidation of compound **4.25** yields the epoxide **4.26**, which can then be attacked by the nucleophilic phenol group leading eventually to the benzofuran ring system **4.27** or the ring system **4.28**.

Problems

Problem 4.4 Suggest a plausible biosynthetic pathway to visnagin **4.29** from compound **4.23**.

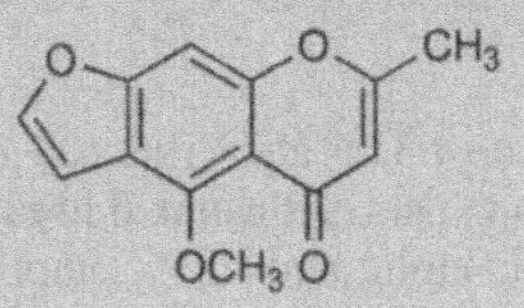

4.29

Problem 4.5 How might acronycine **4.31** originate from compound **4.30**?

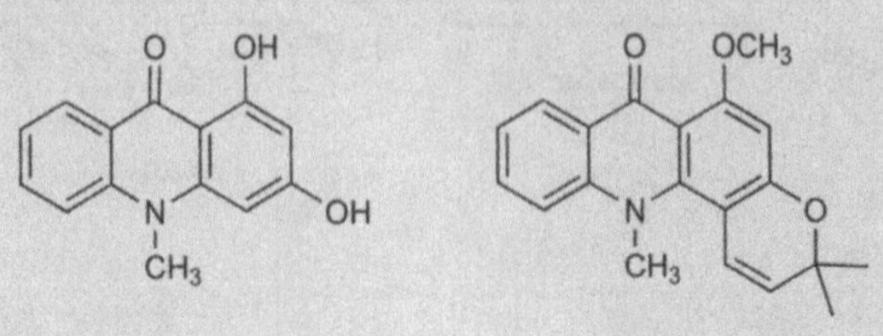

4.30　　　　　　**4.31**

Problem 4.6 How might the compound **4.23** be biosynthesized from the polyketide **4.17**?

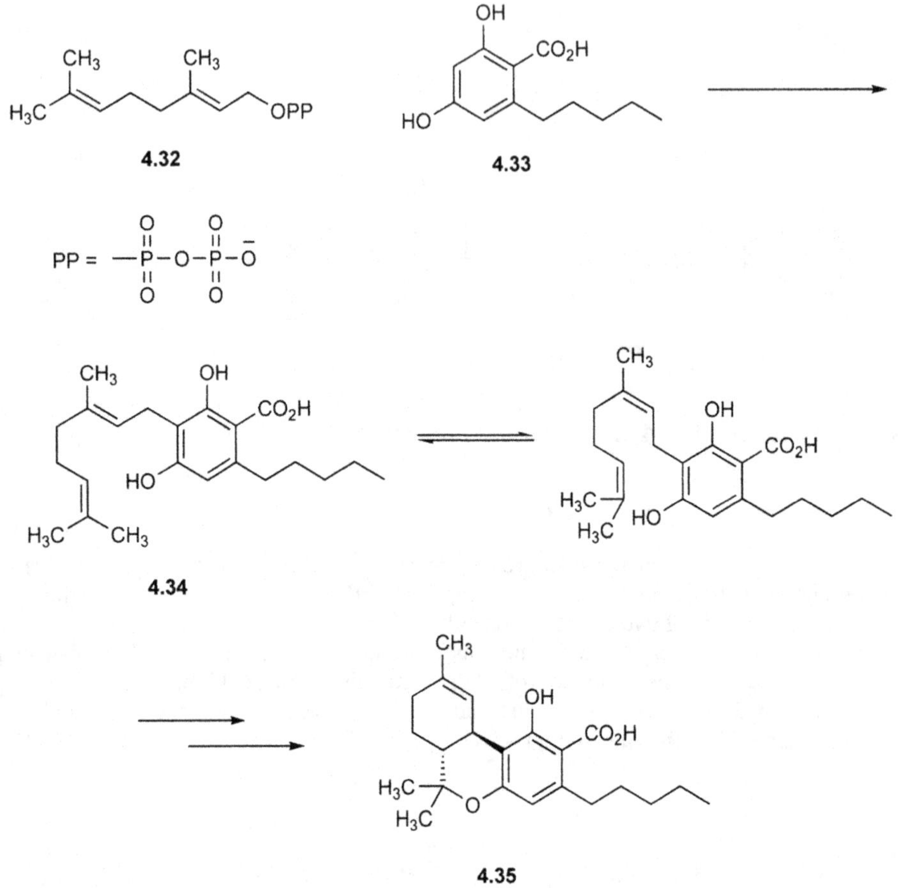

4.32　　　　　　　**4.33**

$$PP = -\overset{\overset{O}{\|}}{\underset{\underset{O}{\|}}{P}}-O-\overset{\overset{O}{\|}}{\underset{\underset{O}{\|}}{P}}-\bar{O}$$

4.34

4.35

Figure 4.17　Biosynthesis of tetrahydrocannabinol **4.35**.

As well as DMAPP **4.22**, geranylpyrophosphate **4.32** is also produced as an intermediate in terpene biosynthesis, and this compound can also act as an alkylating agent. This is illustrated in Figure 4.17 for the formation of cannabigerolic acid **4.34** from geranylpyrophosphate **4.32** and olivetolic acid **4.33**. Compound **4.34** is an intermediate in the biosynthesis of tetrahydrocannabinol **4.35**.

   The natural products orirubenone A **4.37** and C **4.38** have been isolated from the mushroom *Tricholoma orirubens* (Figure 4.18). These two compounds are presumably formed from the geranyl-containing phenolic compound **4.36**. Thus, two allylic oxidations on compound **4.36** yields orirubenone A **4.37** and subsequent cyclization of the acid group onto the alkene group through a conjugate addition reaction gives orirubenone C **4.38**.

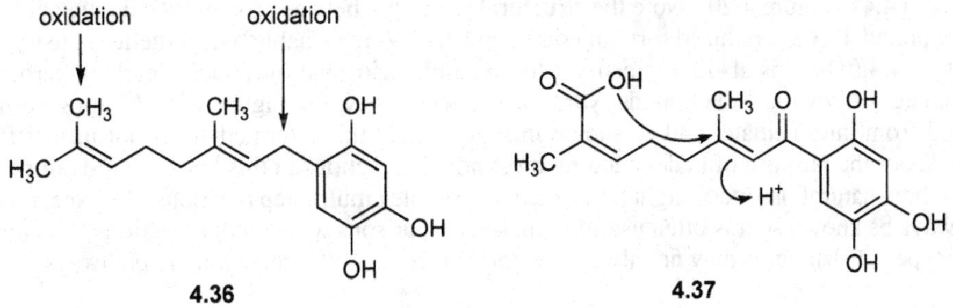

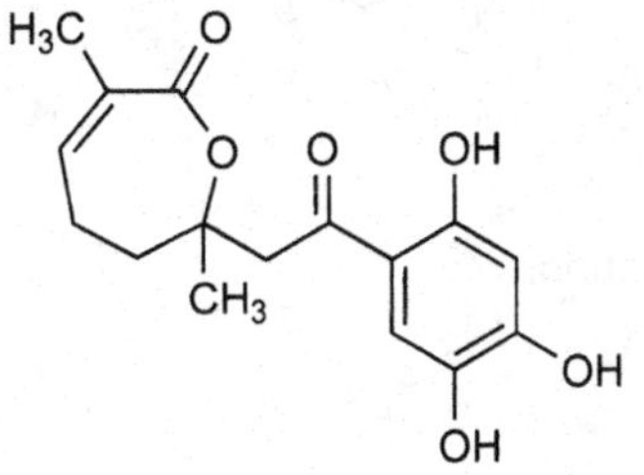

Figure 4.18   Orirubenones A **4.37** and C **4.38**.

4.5.2 Reduction

Ketone **4.39** reduction followed by dehydration is often used as a method for introducing a double bond (Figure 4.19) and we have already met examples of this process in the biosynthesis of unsaturated fatty acids.

An example of the reduction–dehydration sequence shown in Figure 4.19 occurs in the biosynthesis of 6-methylsalicylic acid **4.41** (Figure 4.20). Note the structural similarity between 6-methylsalicylic acid **4.41** and orsellinic acid **4.12**; compound **4.41** is a reduced form of compound **4.12**. A reasonable biosynthetic route to compound **4.41** might use polyketide **4.4** (which is also the precursor to orsellinic acid) and introduce a carbon–carbon double-bond giving the intermediate **4.40** by the reduction–dehydration sequence shown in Figure 4.19. Although 6-methylsalicylic acid **4.41** is formed from intermediate **4.40** as shown in Figure 4.20, this intermediate is not in fact formed directly from polyketide **4.4**. For the purpose of understanding biosynthetic reactions in this book we are only interested in the general concepts of how natural products might be formed. In complex multi-step reactions, the exact order of synthetic events may not always be known so it is often useful to make comparisons with known reactions. You must however be aware that similar types of structures may not always be derived from similar biosynthetic pathways.

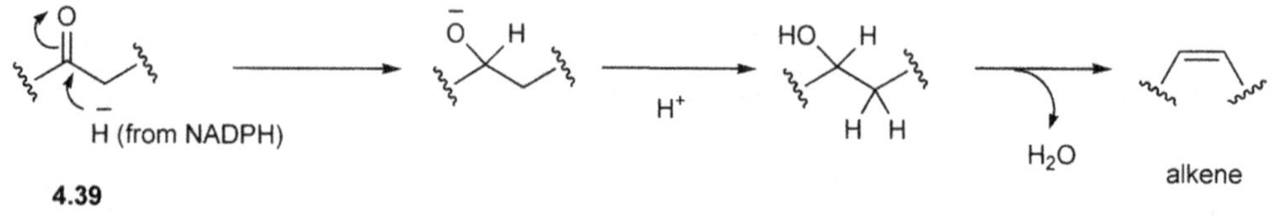

Figure 4.19 Reduction and dehydration.

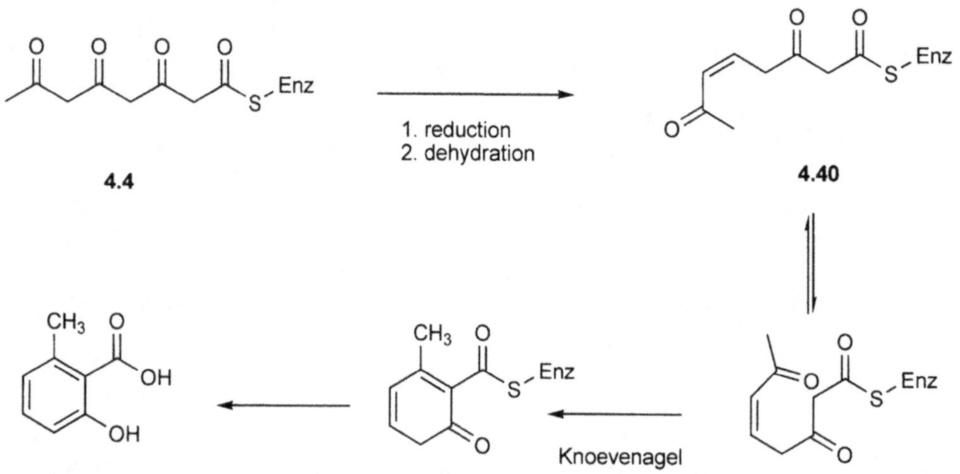

Figure 4.20 Biosynthesis of 6-methylsalicylic acid **4.41**.

Problems

Problem 4.7 The alkene **4.40** is in fact biosynthesized from the polyketide **4.3** (Figure 4.2). Propose a reasonable synthesis of compound **4.40** from polyketide **4.3** which *does not* involve the polyketide **4.4** as an intermediate. You may want to re-read section 3, sub-section 3.5.

Problem 4.8 Wailupemycin G **4.42** is produced by the bacterium *Strepomyces maritimus* from the starter unit, benzoyl coenzyme A (PhCOSCoA) and 7 molecules of malonyl coenzyme A. Deduce a plausible biosynthetic pathway to wailupemycin G **4.42**.

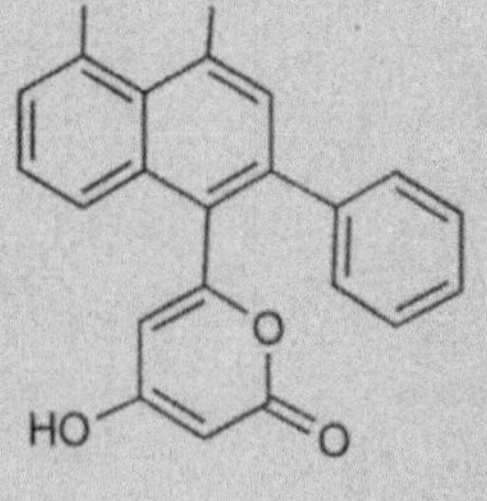

4.42

Problem 4.9 The polyketide **4.43** and compound **4.44** are proposed intermediates in the biosynthesis of isoelaeocarpiline **4.45**, an alkaloid isolated from the Australian rainforest tree, *Elaeocarpus grandis*. Show how polyketide **4.43** might be transformed into compound **4.44** and also propose a reasonable mechanism for the conversion of compound **4.44** into isoelaeocarpiline **4.45**.

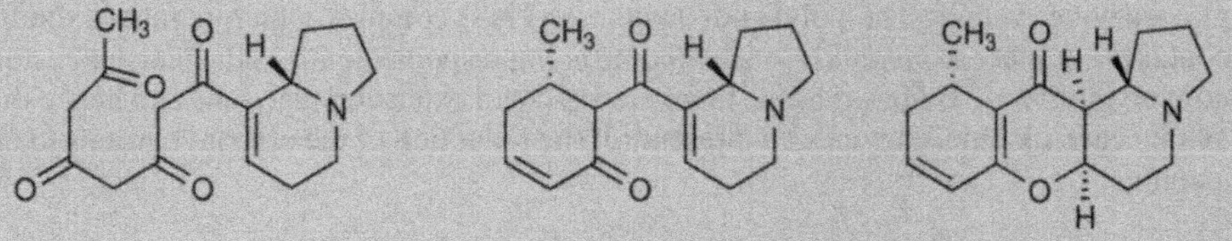

4.43   **4.44**   **4.45**

Problem 4.10 Suggest a reasonable biosynthetic pathway to climacostol **4.46**, a defence toxin of the heterotrich ciliate *Climacostomum virens*. You will need a decarboxylation reaction in this biosynthesis (see section 4.5.4).

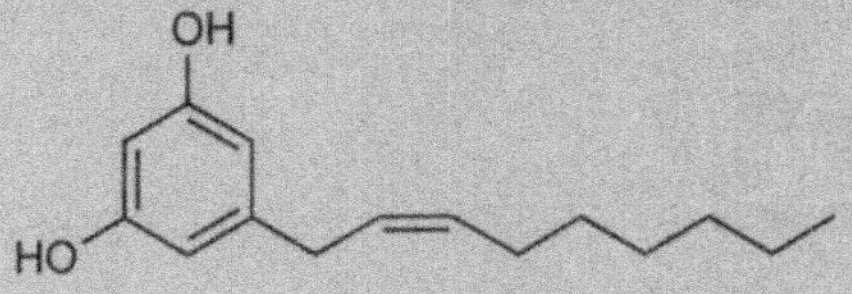

4.46

The polyketide pathway has furnished a diverse range of lactones and some examples of this class of compound has already been discussed in section 3, sub-section 3.6. Some δ-lactones such as prelactone B **4.47** and C **4.48** can be formed from appropriate polyketide precursors as outlined in Figure 4.21. During the formation of these lactones, the carbonyl groups in the polyketide chain have been reduced to alcohols.

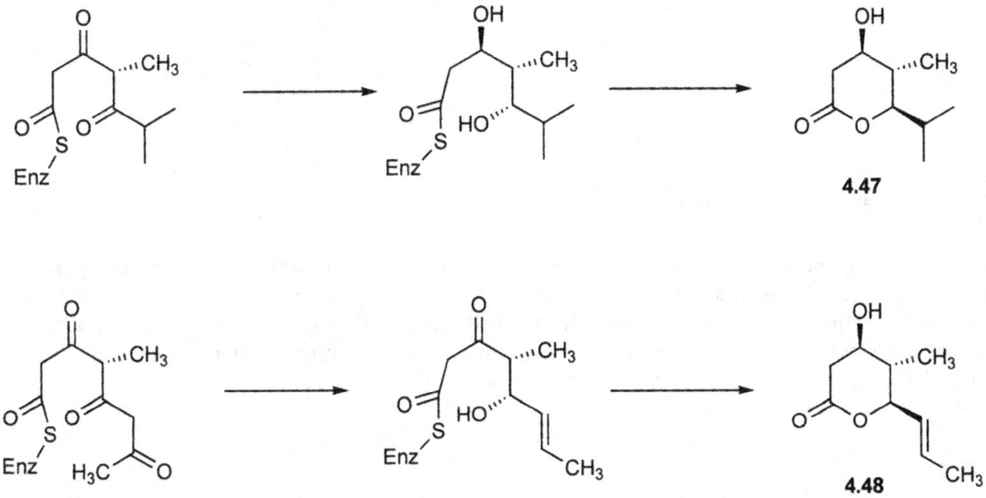

Figure 4.21 Formation of prelactones B **4.47** and C **4.48**.

Nature also produces a diverse range of macrocyclic lactones. One important example is 6-deoxyerythronolide B **4.50** which is the precursor to the erythromycin class of antibiotics. Compound **4.50** can be formally derived from a propionate starter unit and six molecules of methylmalonyl coenzyme A as illustrated in Figure 4.22. Although it is convenient to compare the relationship between the carbonyl and alcohol groups in polyketide **4.49** and 6-deoxyerythronolide B **4.50**, in reality the reduction of the carbonyl groups occurs as the carbon chain is extended by sequential addition of the methylmalonyl coenzyme A units. The polyketide synthase (PKS) complex which regulates the formation of the polyketide chain contains a number of *modules* – each module being responsible for each chain extension (Figure 4.23). In the case of 6-deoxyerythronolide B biosynthesis, there are six chain extension reactions and hence the PKS complex has six domains. Within each module, enzymes are present for the reduction of the carbonyl groups to either methylene ($>CH_2$) or to an alcohol.

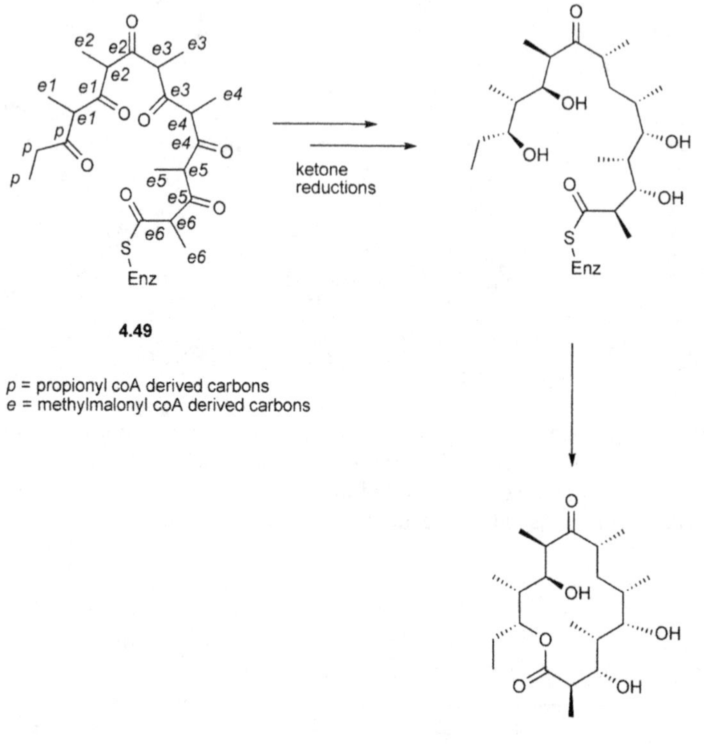

p = propionyl coA derived carbons
e = methylmalonyl coA derived carbons

Figure 4.22   6-Deoxyerythronolide B **4.50**.

The PKS complex can be genetically modified to produce a variety of *unnatural* lactones. For example, if biosynthesis beyond module 2 is prevented, then release of the polyketide chain at this point will yield the lactone **4.51** (Figure 4.24). A second example of manipulation of the PKS complex is illustrated by the formation of the lactone **4.52** (Figure 4.25). If reduction of the carbonyl group in module 5 is prevented, this carbonyl group eventually appears in the product **4.52**, whereas normally an alcohol group would be present.

Problem

Problem 4.11 Suggest a reasonable biosynthesis of 10-deoxymethynolide **4.53**, a macrocyclic lactone produced by the organism *Strepomyces venezuelae*.

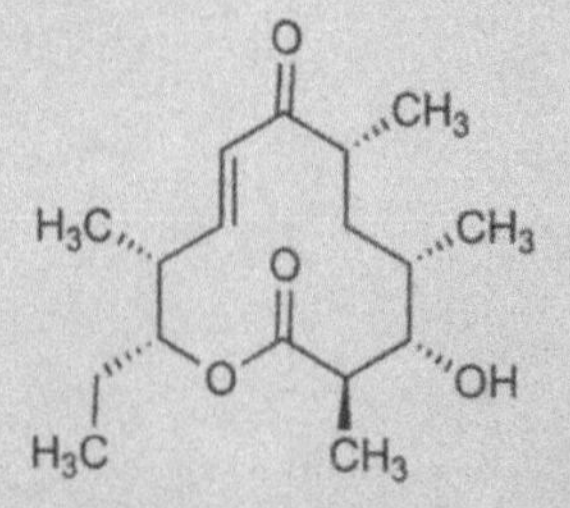

4.53

4.50

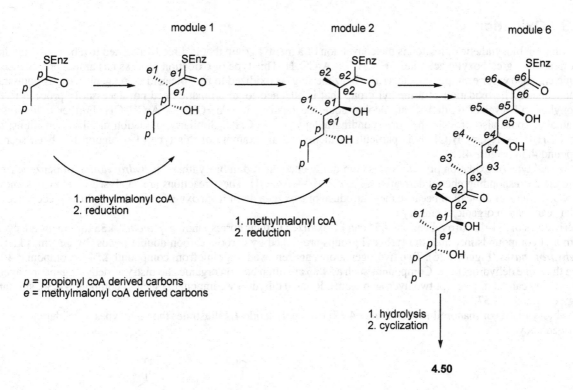

Figure 4.23 Chain extension reactions in 6-deoxyerythronolide B **4.50** biosynthesis.

4.51

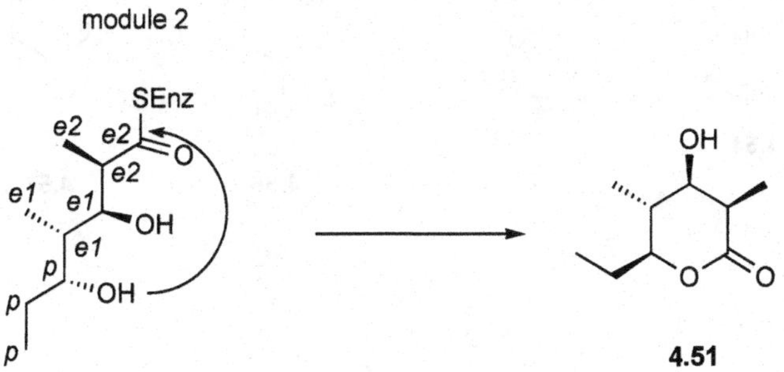

Figure 4.24 Formation of lactone **4.51**.

4.52

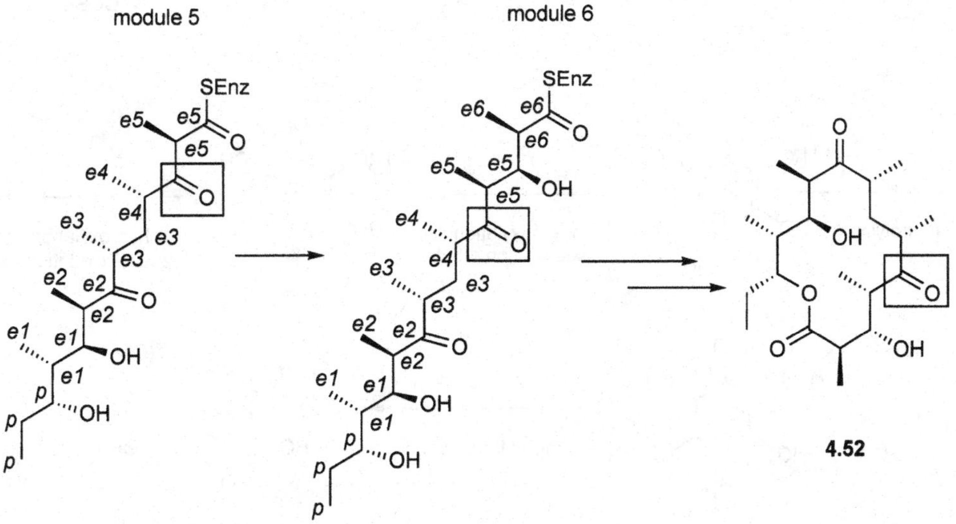

Figure 4.25 Biosynthesis of lactone **4.52**.

4.5.3 Oxidation

One common biosynthetic oxidation is the conversion of a methyl group that is directly attached to a benzene ring into its corresponding carboxylic acid, i.e. $Ar-CH_3 \rightarrow ArCO_2H$. This type of oxidation process has ample precedence in organic chemistry; for example toluene (Ar = benzene ring) is oxidized to benzoic acid by reagents such as potassium permanganate. The oxidation of a methyl group that is attached to an aromatic need not necessarily proceed to the carboxylic acid level of oxidation; an alcohol could be formed, i.e. $Ar-CH_3 \rightarrow ArCH_2OH$. Further oxidation of this alcohol could also provide the corresponding aldehyde, $Ar-CHO$. Similarly, oxidation at allylic positions, i.e. $R_2C{=}CCH_2R \rightarrow R_2C{=}CH(OH)R$ is particularly facile and an example of this type of oxidation has been seen in compound **4.36** (Figure 4.18).

A second type of oxidation process that is commonly encountered in biosynthesis is *hydroxylation* of benzene rings giving the corresponding phenol derivatives, i.e. $Ar-H \rightarrow Ar-OH$. These reactions are catalysed by enzymes known as *monooxygenases* (so-called because they introduce one oxygen atom from oxygen) and there is little precedence for similar reactions in organic chemistry.

Derivatives of 1,4-dihydroxybenzene **4.54** can be oxidized to their corresponding *quinones* **4.55** (quinones are cyclic dicarbonyl compounds in which the carbonyl groups are linked by carbon–carbon double bonds) by enzymes known as *dehydrogenases* (Figure 4.26). Two hydrogen atoms are removed in going from compound **4.54** to compound **4.55**, hence the name dehydrogenase. Compounds such as **4.55** are often used in organic chemistry as dehydrogenation agents because they can readily accept two hydrogen atoms. Related dihydroxy compounds, such as compound **4.56**, similarly can give quinones **4.57**.

The biosynthesis of shanorellin **4.58** (Figure 4.27) from polyketide **4.4** illustrates the three types of oxidation process described above.

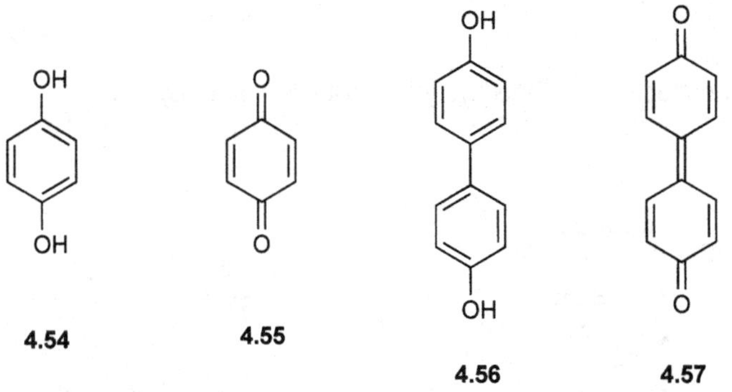

Figure 4.26 Quinones.

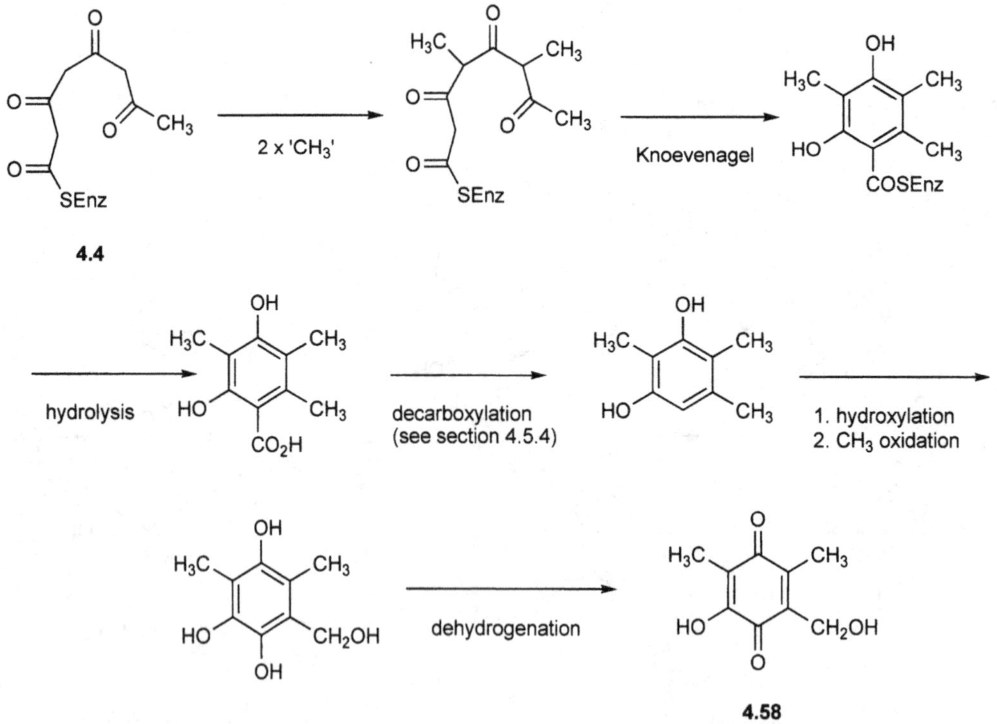

Figure 4.27 Biosynthesis of shanorellin **4.58**.

Condensation of propionyl coenzyme A with nine malonyl coenzyme A units affords the polyketide **4.65** which eventually furnishes the quinone-containing compound **4.66** (Figure 4.28). Note that one of the polyketide carbonyl groups is reduced to a methylene unit, and that the second quinone oxygen atom is introduced by oxidation. Methylation and cyclization of compound **4.66** then gives aklaviketone **4.67** which is an important biosynthetic intermediate in the anthracycline class of antibiotics.

The last type of oxidation process we need to cover is called *phenolic coupling* (Figure 4.29). In this reaction, phenols (i.e. hydroxybenzenes) are oxidized to their corresponding *phenoxy radicals* **4.68**. The unpaired electron in a phenoxy radical can be delocalized over the oxygen atom, the carbon atom at the 2-position and the carbon atom at the 4-position, as shown in Figure 4.29. Once phenoxy radicals have been generated, they dimerize by pairing the unpaired electron of one phenoxy radical with the unpaired electron of a second phenoxy radical. One example of this dimerization process is shown in Figure 4.29, in which the carbon atom at the 2-position of one phenoxy radical becomes bonded to the carbon atom at the 4-position of a second phenoxy radical.

There is ample precedence for phenoxy radical dimerisations in organic chemistry, phenols (or their corresponding anions, Ar—O$^-$) are readily oxidized to phenoxy radicals by numerous mild oxidizing agents, for example potassium ferricyanide.

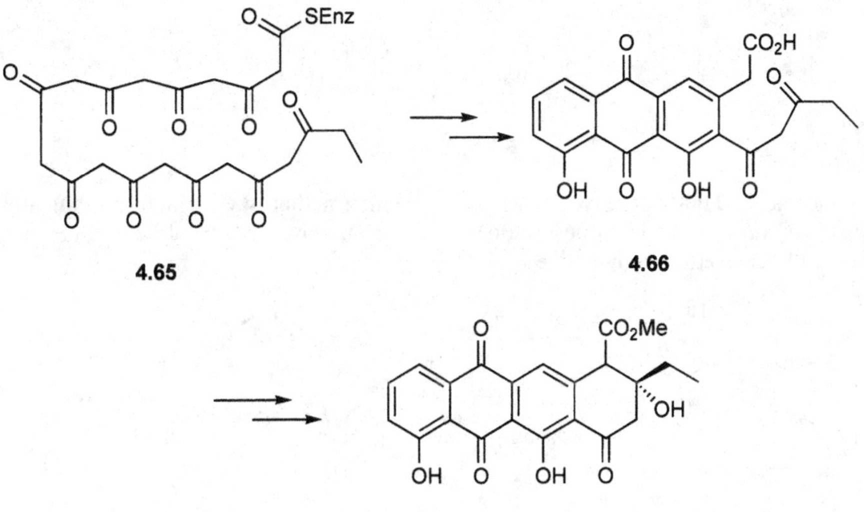

Figure 4.28   Biosynthesis of aklaviketone **4.67**.

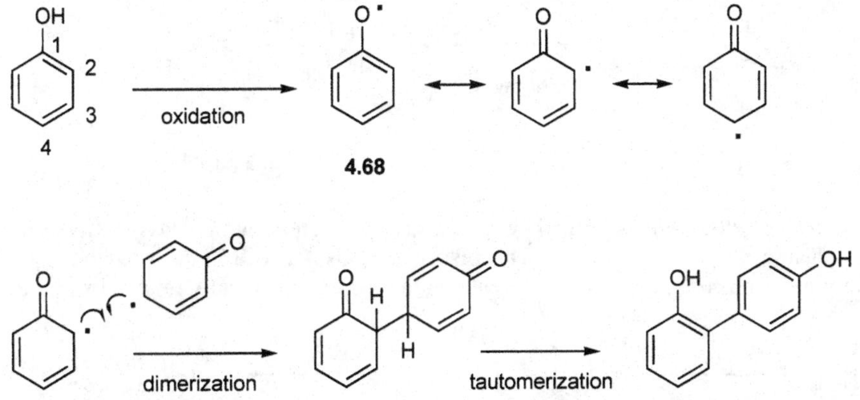

Figure 4.29   Phenoxy radicals.

53

Problems

Problem 4.12 The naphthalene derivative **4.59** forms part of the structure of the anti-tumor antibiotic azinomycin A **4.60**. Propose a reasonable biosynthesis of compound **4.59** from the polyketide **4.61**.

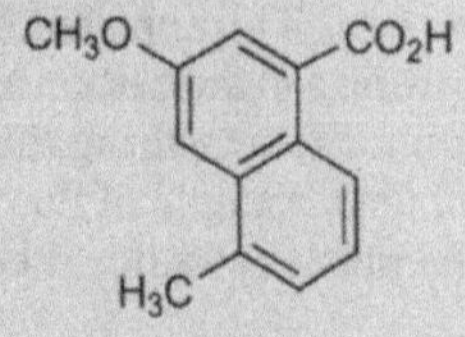

4.59

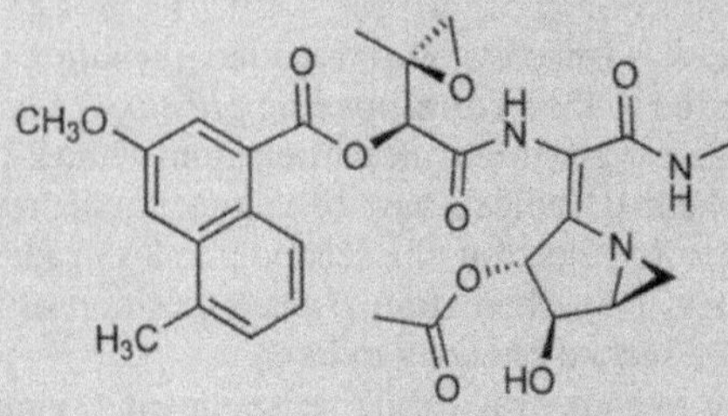

4.60

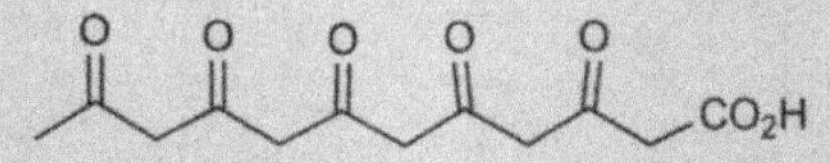

4.61

Problem 4.13 The lactone **4.62** undergoes oxidation giving the intermediate **4.63** which subsequently cyclizes yielding, after methylation, monocerin **4.64**. Propose a polyketide precursor to lactone **4.62** and demonstrate how your chosen precursor might be converted into lactone **4.62**.

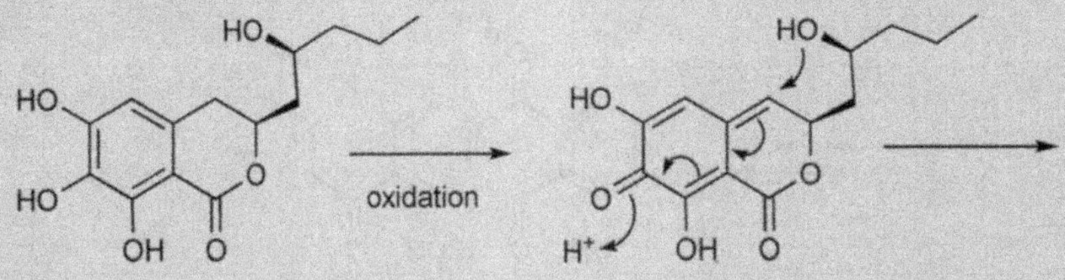

4.62

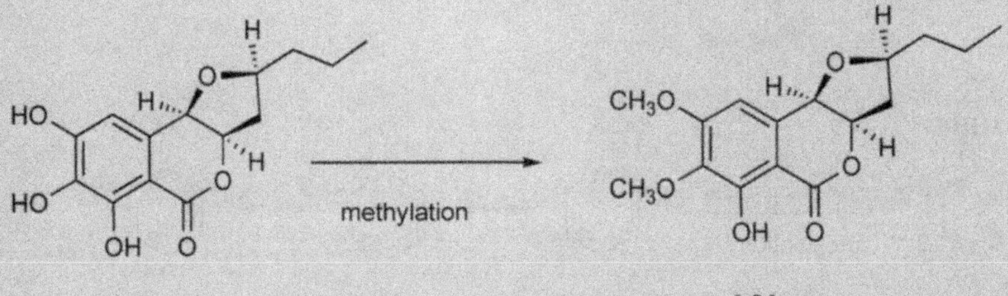

4.63

4.64

Problem 4.14 Pummerer's ketone **4.69** is formed by oxidation of *para* cresol with potassium ferricyanide under basic conditions as shown below. Propose reasonable mechanisms for steps A and B in this reaction. This reaction can be considered as a biomimetic reaction, i.e. an organic reaction that is mechanistically related to a biological reaction.

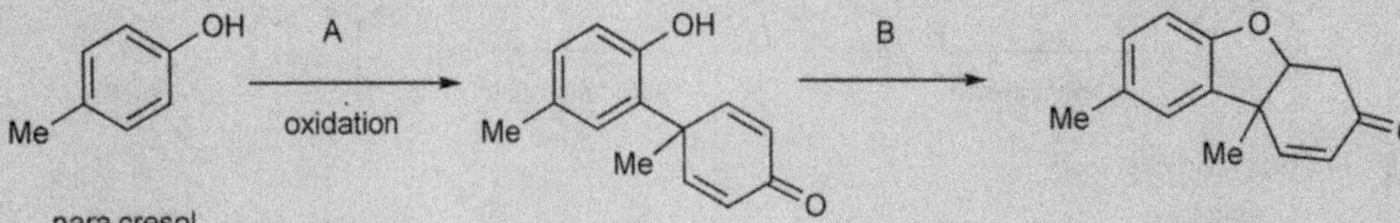

para cresol

4.69

Problem 4.15 Suggest a mechanism for the conversion of the phenol **4.19** into usnic acid **4.20**.

Problem 4.16 Propose a suitable polyketide precursor for the phenol **4.70** and show how cyclization of your chosen precursor might give compound **4.70**. There are two possible modes of cyclization.

The biosynthesis of griseofulvin **4.71**, a natural product with fungicidal activity isolated from the mould *Penicillium griseofulvum*, proceeds from the phenolic intermediate **4.70** as shown in Figure 4.30. The chlorine atom is presumably introduced as Cl$^+$ in an electrophilic aromatic substitution reaction; Cl$^+$ can be produced by oxidation of chloride (Cl$^-$).

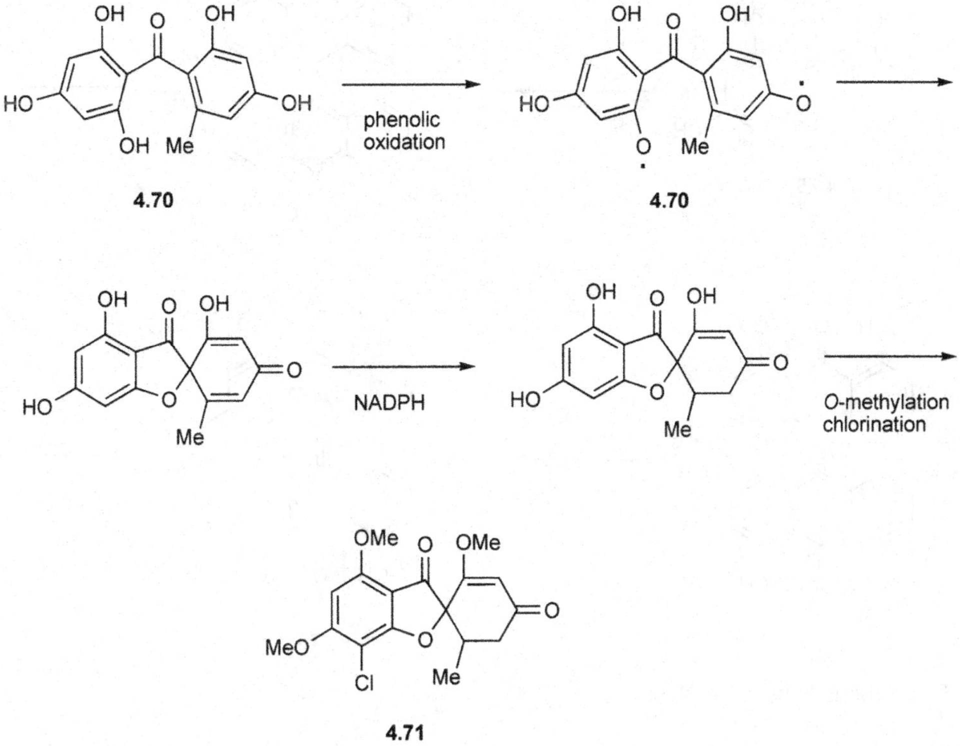

Figure 4.30 Biosynthesis of griseofulvin **4.71**.

Problem

Problem 4.17 Show how the phenolic compound **4.73** might be derived from benzoyl coenzyme A **4.72** and three units of malonyl coenzyme A, and then demonstrate how compound **4.73** might be converted into gentisein **4.74**.

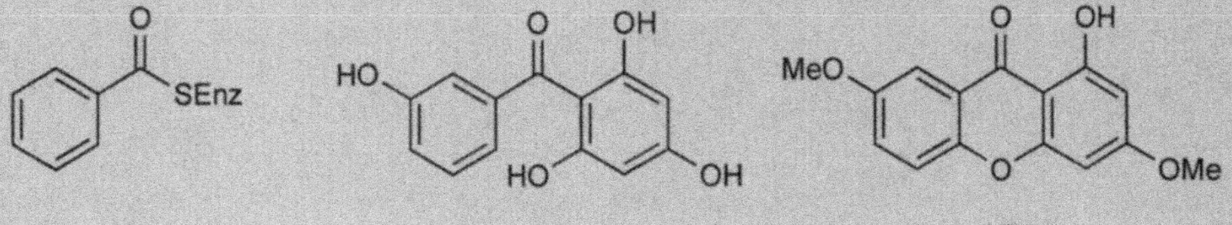

Figure 4.31 shows the biosynthesis of the pigment, hypericin **4.76** from the precursor **4.75**. Three phenoxy radical dimerizations and one quinone forming oxidation are involved in this biosynthesis.

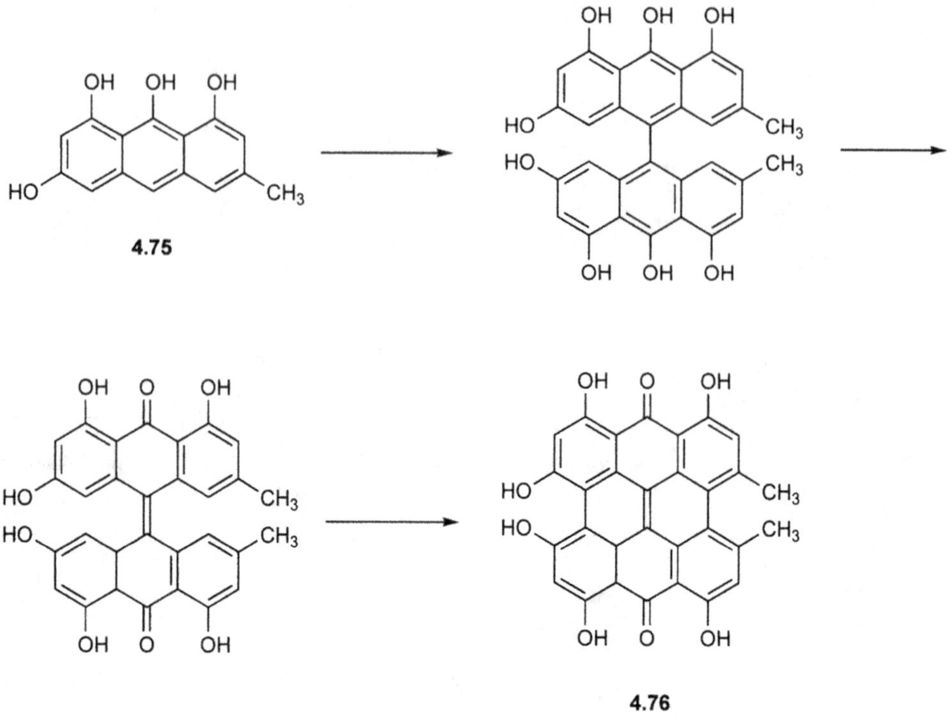

Figure 4.31 Biosynthesis of hypericin **4.76**.

Problem

Problem 4.18 Suggest a biosynthetic pathway to compound **4.75** from the polyketide **4.77**. You will need a decarboxylation reaction in this biosynthesis (see section 4.5.4).

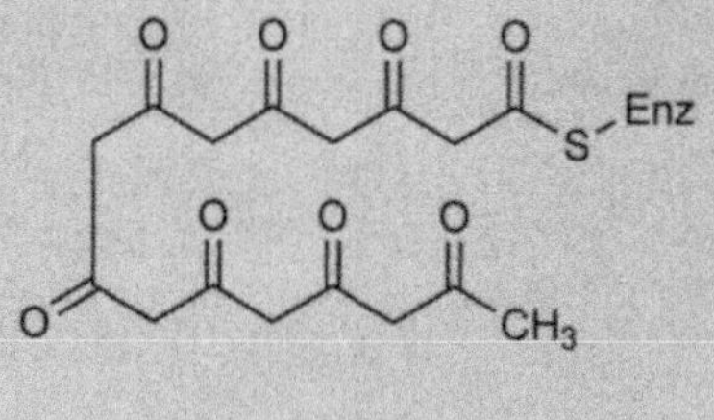

4.5.4 Decarboxylation

Decarboxylation occurs readily in biosynthetic and organic reactions particularly in 2-hydroxybenzoic acid derivatives (Figure 4.32). Note the β-keto-acid fragment as an intermediate which is highly susceptible to decarboxylation. The biosynthetic routes in problems 4.10 and 4.18 use a similar decarboxylation mechanism.

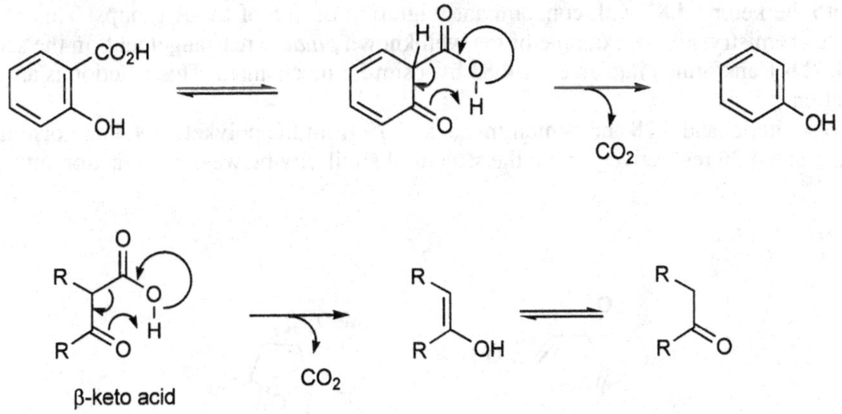

Figure 4.32 Decarboxylation.

4.5.5 Modifications to the Carbon Skeleton

The products produced by polyketide cyclization reactions can often undergo modifications to their carbon skeletons thus further extending the range of this genre of natural products. Stipitatic acid **4.78** and penicillinic acid **4.79** (Figure 4.33) are two representative examples. Before we consider how the carbon skeletons of these two natural products are formed it is useful to consider the two rearrangement reactions depicted in Figure 4.34. In the first reaction, the diol **4.80** is converted into the ketone **4.81** with concomitant migration of one of the R groups. This reaction has ample precedence in organic chemistry, it is an example of the well known *pinacol* rearrangement. In the second reaction the peroxy compound **4.82** is transformed into an ester **4.83** by a similar mechanism. This reaction is an application of the *Baeyer–Villiger* reaction.

Plausible routes to stipitatic acid **4.78** and penicillinic acid **4.79** from the polyketide **4.4** and orsellinic acid **4.12** are shown in Figures 4.35 and 4.36 respectively. Note the structural similarity between the reaction intermediates in these two reactions.

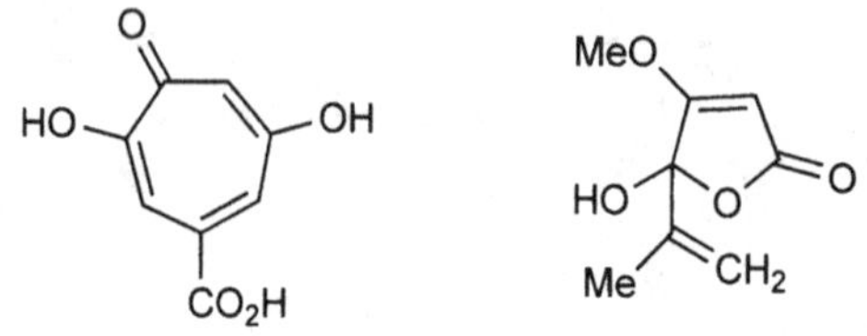

Figure 4.33 Stipitatic acid **4.78** and penicillinic acid **4.79**.

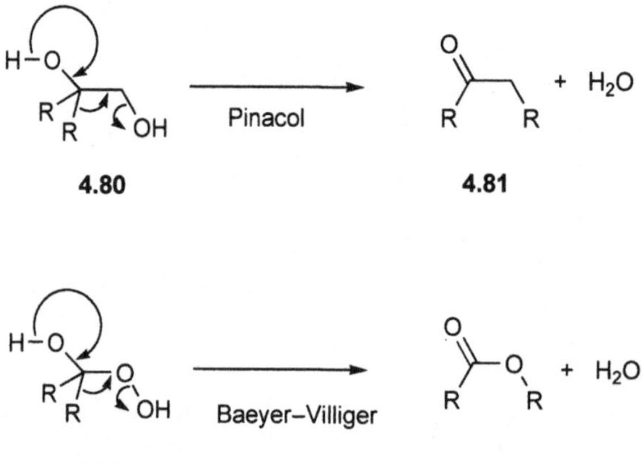

Figure 4.34 Pinacol and Baeyer–Villiger reactions.

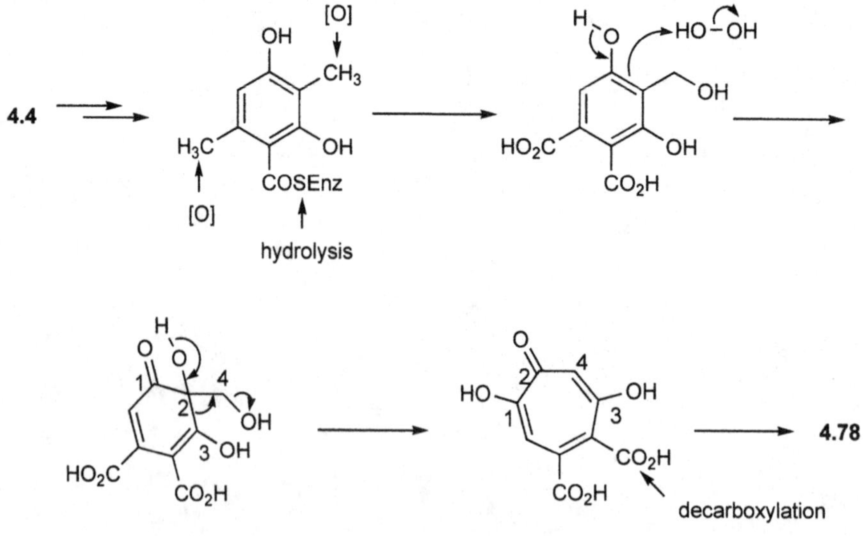

Figure 4.35 Biosynthesis of stipitatic acid **4.78**.

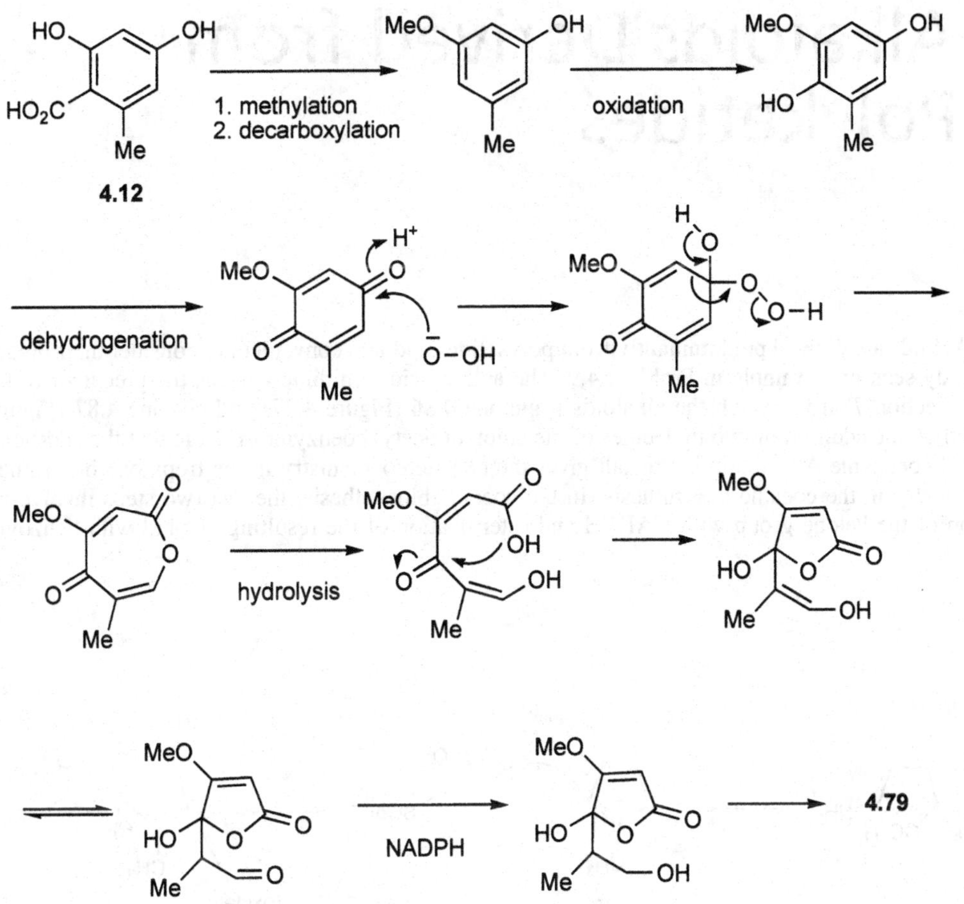

Figure 4.36 Biosynthesis of penicillinic acid **4.79**.

4.6 Alkaloids Derived from Polyketides

Numerous alkaloids are derived predominantly from polyketides and it is convenient to consider their biogenesis here. We have already seen one example in Problem 4.9. The amino acid, ornithine **4.84**, is the precursor of the iminium salt **4.85** (see section 7) from which the alkaloids tropinone **4.86** (Figure 4.37) and cocaine **4.87** (Figure 4.38) are biosynthesized by the addition of two molecules of the anion of acetyl coenzyme A. Note that the addition of the first anion of acetyl coenzyme A to the iminium salt gives the (*R*)-stereochemistry in the tropinone biosynthesis and the (*S*)-stereochemistry in the cocaine biosynthesis. In the cocaine biosynthesis, the last two steps involve a stereospecific reduction of the ketone group with NADPH and esterification of the resulting alcohol with benzoyl coenzyme A **4.72**.

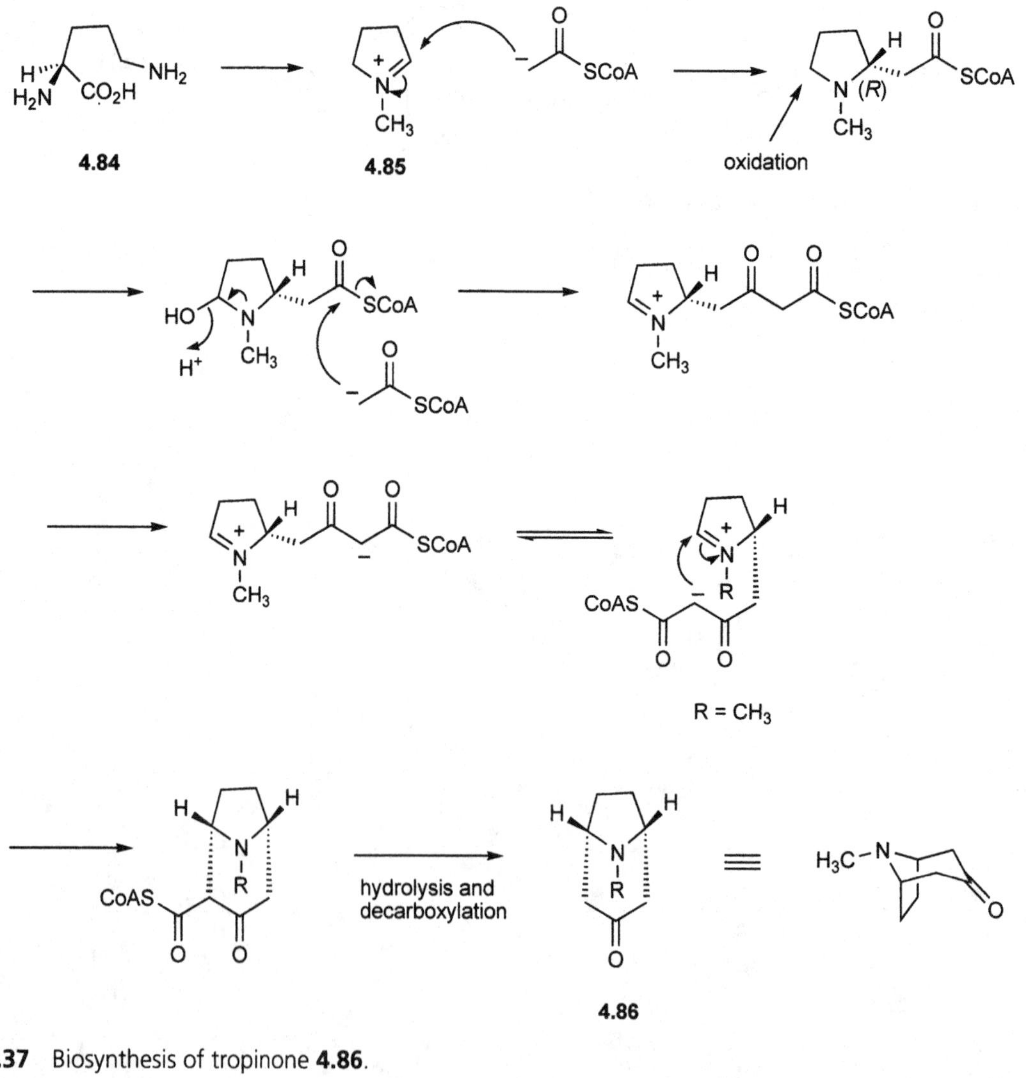

Figure 4.37 Biosynthesis of tropinone **4.86**.

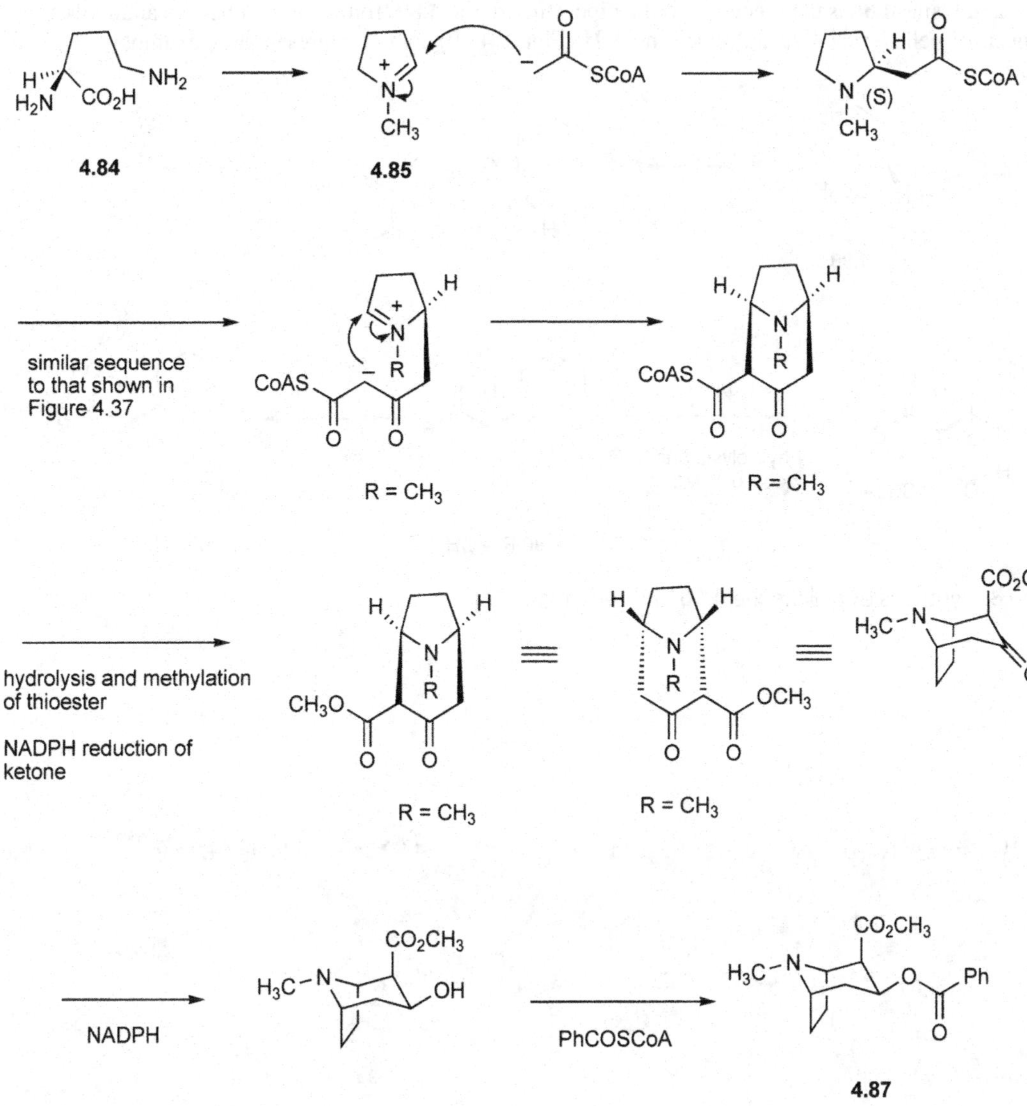

Figure 4.38 Biosynthesis of cocaine **4.87**.

Problem

Problem 4.19 Robinson showed that racemic tropinone **4.86** (one enantiomer shown) could be prepared from the components depicted below under virtually physiological conditions in a *biomimetic* reaction. Propose a reasonable mechanism for this reaction.

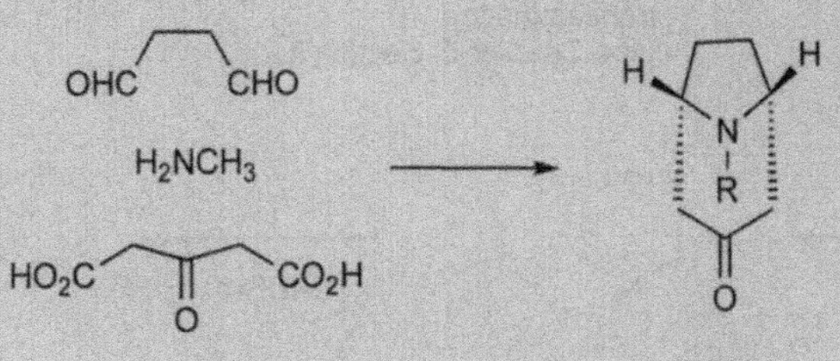

Alkaloids Derived from Polyketides

The amino acid lysine **4.88** is the precursor to the piperidinium ion **4.89** from which numerous alkaloids are derived. The biosynthesis of pelletierine **4.90** and sedamine **4.91** (Figure 4.39) are two representative examples.

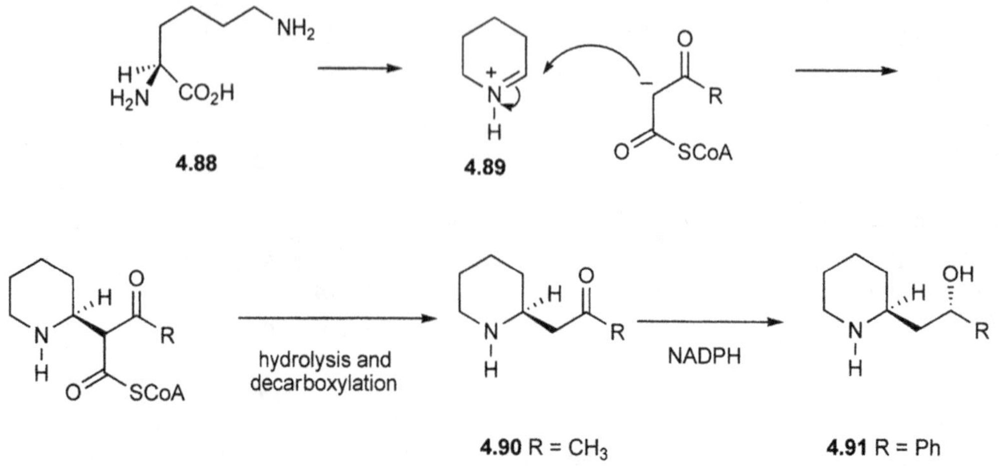

Figure 4.39 Biosynthesis of pelletierine **4.90** and sedamine **4.91**.

Problem

Problem 4.20 Propose reasonable biosynthetic pathways to anaferine **4.92** and lobelanine **4.93** from compound **4.89**.

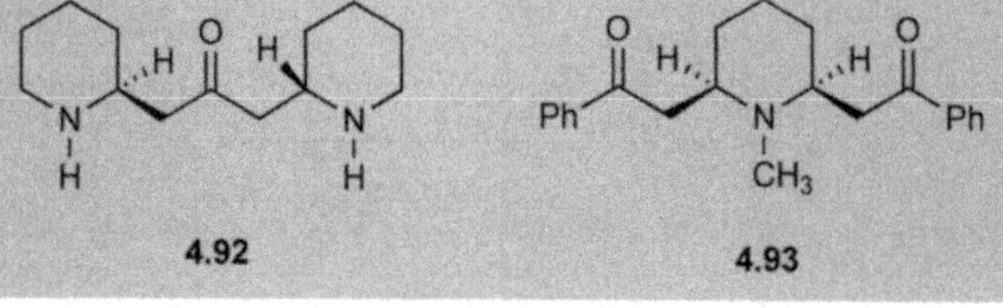

The poisonous hemlock plant produces several alkaloids including coniine **4.94**. At first glance coniine might be considered to have the same biosynthetic origin as the structurally related pelletierine **4.90**, but it is in fact produced as outlined in Figure 4.40.

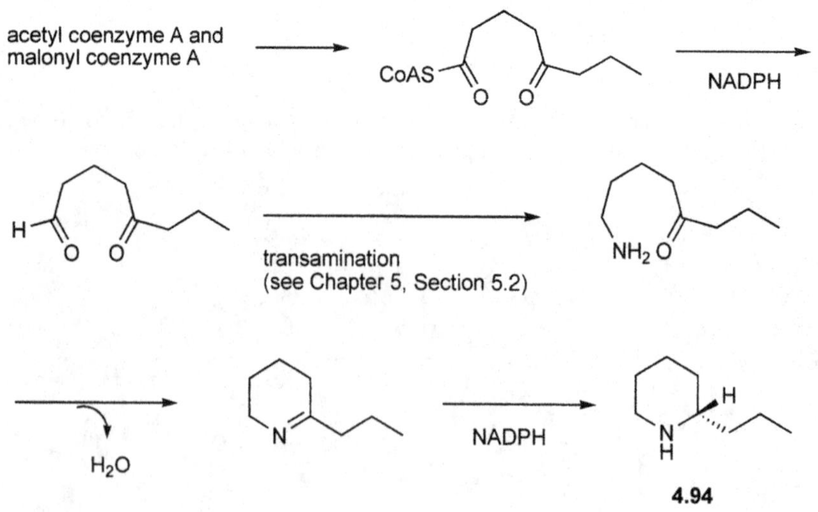

Figure 4.40 Biosynthesis of coniine **4.94**.

The coenzyme A derivative **4.95** of anthranilic acid can act as a starter unit in the formation of polyketide chains. This is illustrated in Figure 4.41 for the formation of the alkaloid arborinine **4.96**.

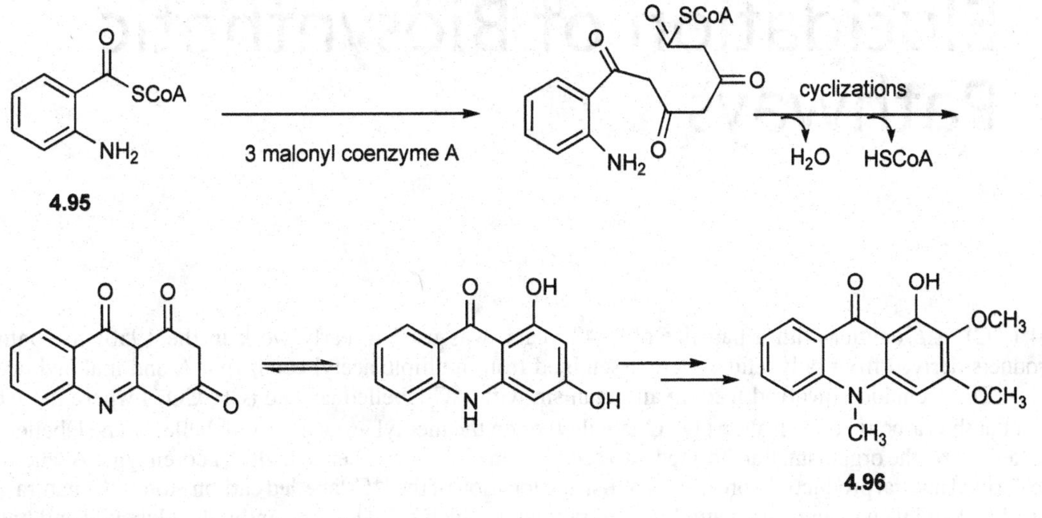

Figure 4.41 Biosynthesis of arborinine **4.96**.

4.7 The Use of Isotopes in the Elucidation of Biosynthetic Pathways

Carbon-14 (^{14}C), a β-emitter with a half-life of 5640 years, was used in early work in the 1950s to confirm that natural products derived from polyketides were assembled from multiple acetyl coenzyme A and malonyl coenzyme A molecules. The technique involved feeding an organism with ^{14}C-labelled acetate ($CH_3CO_2^-$) where the label was located either at the carbonyl carbon atom (1-^{14}C-labelled) or on the methyl group (2-^{14}C-labelled). The labelled acetate was then taken-up by the organism, transformed into acetyl coenzyme A and hence malonyl co-enzyme A, and then into polyketide derived natural products. In order to establish the location of the ^{14}C-labelled carbon atoms, the natural product would be isolated and then chemically degraded into fragments which contained the individual labelled and unlabelled carbon atoms. For example, feeding the mould *Penicillium patulum* with 1-^{14}C-labelled acetate gave a polyketide **4.4** (see Figure 4.21) with the ^{14}C label located at each of the carbonyl carbon atoms. Polyketide **4.4** is then transformed into 6-methylsalicylic acid **4.41** with the ^{14}C labels located at the carboxyl-carbon atom and positions 2, 4 and 6 of the benzene ring. Degradation of 6-methylsalicylic acid **4.41** by decarboxylation would yield labelled CO_2. Further degradation of the aromatic ring confirmed the positions of the other labelled carbon atoms. An obvious disadvantage of using ^{14}C-labelled acetate is that relatively large quantities of the natural product are required for chemical degradation.

During the 1970s the elucidation of biosynthetic pathways was revolutionized by the introduction of carbon-13 (^{13}C) isotopes and nuclear magnetic resonance (NMR) spectroscopy. The natural abundance of ^{13}C is 1.1% and thus *feeding* organisms with ^{13}C-labelled acetate gives an NMR spectrum in which some of the resonances will have been enhanced. If the natural abundance ^{13}C-NMR spectrum of a natural product has been assigned, then the positions at which labels have been incorporated will be apparent because of the enhanced signals. One great advantage of this technique is that only relatively small quantities of the natural product are required to obtain the ^{13}C-NMR spectrum and chemical degradation is not necessary. Thus, if 1-^{13}C-labelled acetate was used instead of 1-^{14}C acetate in the 6-methylsalicylic acid **4.41** (Figure 4.20), the ^{13}C-NMR spectrum of compound **4.41** would show enhanced signals at the carboxyl-carbon atom and carbons 2, 4 and 6 of the aromatic ring.

The 1,2-^{13}C doubly-labelled acetate has been used extensively and to great effect in determining biosynthetic pathways. In this technique, all the acetate-derived carbon atoms are labelled and adjacent nuclei (C_1 and C_2) will exhibit $^{13}C_1$–$^{13}C_2$ coupling ($J_{1,2}$), and thus these adjacent nuclei will be observed as doublets in the NMR spectrum. Adjacent nuclei can therefore be identified through their coupling constants. When consecutive acetyl coenzyme A/malonyl coenzyme A units are joined, this will give a sequence of 4 carbon atoms: C_1–C_2–C_3–C_4. Since the incorporation of labelled isotopes in feeding experiments is usually quite low, statistically it is unlikely that these labelled units will be incorporated sequentially and hence coupling is not observed between C_2 and C_3. Thus, for the sequence C_1–C_2–C_3–C_4 we would expect to see two coupling constants $J_{1,2}$ and $J_{3,4}$ resulting from the two component acetyl coenzyme A/malonyl coenzyme A precursors. Returning to the methylsalicylic acid **4.41** example we would expect to observe the pattern depicted in Figure 4.42 if [1,2-^{13}C]-acetate was used in a feeding experiment. The bold lines in Figure 4.42 are used to designate adjacent labelled nuclei.

With large polyketides, there are often several ways in which the polyketide chain can be folded prior to cyclization. Different folding modes can often be identified by double-labelling experiments. An illustrative example is the biosynthesis of rubrofusarin **4.97** from the polyketide **4.14** (Figure 4.43). Each folding mode yields a different pattern of isotope incorporation, with the top folding mode being consistent with the experimentally observed coupling data.

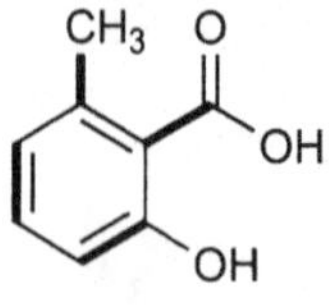

4.41

Figure 4.42 Double ^{13}C labelling of 6-methylsalicylic acid **4.41**.

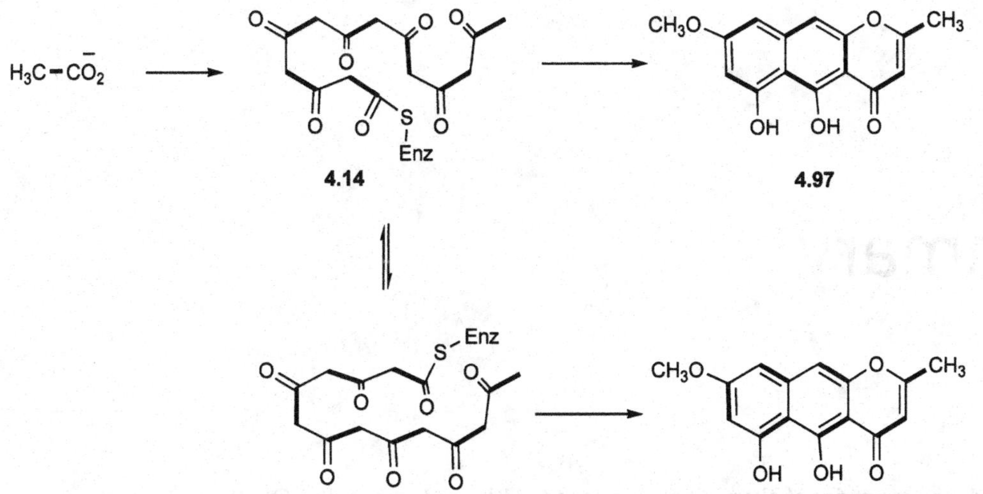

Figure 4.43 Elucidation of polyketide folding modes in rubrofusarin **4.94**.

The above examples illustrate briefly how isotopes have been used to establish biosynthetic pathways. Many other isotopes have been used in labelling studies e.g. deuterium, ^{15}N and ^{18}O, as well as combinations of different isotopes within a substrate. In addition to acetate, other labelled precursors e.g. labelled propionate can be utilized in feeding experiments. Labelled methionine, in which the S-methyl group has been labelled yields methyl-labelled S-adenosyl methionine, and hence the origin of methyl groups in methylation reactions can be traced.

Summary

(a) Polyketides have a series of alternating methylene ($>CH_2$) and carbonyl ($>C{=}O$) groups.
(b) Polyketides undergo cyclization reactions of the Knoevenagel and Claisen types giving phenolic products.
(c) Secondary modifications (alkylations, decarboxylations, oxidations, reductions) to polyketides or their cyclized products are common.
(d) Many alkaloids are of substantial polyketide origin.
(e) Isotopes can be used to elucidate biosynthetic pathways.

The Shikimic Acid Pathway

Objectives

After studying this section you will be able to:

(a) Recognize the role of shikimic acid, chorismic acid and prephenic acid in biosynthesis.
(b) Describe how the amino acids tryptophan, tyrosine and phenylalanine are biosynthesized.
(c) Describe how cinnamic and benzoic acid derivatives are biosynthesized from phenylalanine and recognize that cinnamic and benzoic acid derivatives are precursors to other natural products.

5.1 Introduction

Shikimic acid **5.1** (Figure 5.1) is the precursor to many natural products that contain an aromatic benzene ring as part of their structure. There are two other important intermediates in the synthesis of these natural products and they are *chorismic acid* **5.2** and *prephenic acid* **5.3** (Figure 5.1). Chorismic acid is biosynthesized from shikimic acid. Prephenic acid is derived from chorismic acid by a *rearrangement* (note that these two compounds **5.2** and **5.3** have the same empirical formulae). This section will describe how chorismic acid **5.2** and prephenic acid **5.3** are used in biosynthesis to form a variety of natural products containing a benzene ring. Note that many *phenolic* natural products (i.e. hydroxybenzene derivatives) are biosynthesized *via* the polyketide pathway and some phenolic compounds can be derived either from polyketides or shikimic acid depending upon the organism in which biosynthesis occurs. Phenolic products occur quite commonly as glycosides in which one (or more) of the phenolic hydroxyl groups is attached to a sugar residue. For simplicity, only the phenolic parent structures are discussed in this section.

In this section we will see how chorismic acid **5.2** is the precursor of anthranilic acid **5.4** and of the amino acid tryptophan **5.5**, and also how prephenic acid **5.3** is the precursor to the amino acids phenylalanine **5.6** and tyrosine **5.7** (Figure 5.2). Note that all of these compounds contain a benzene ring; this benzene ring was originally the six-membered

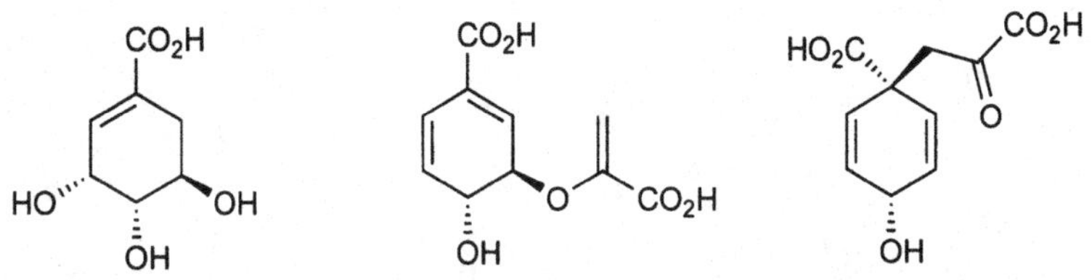

Figure 5.1   Shikimic acid **5.1**, chorismic acid **5.2** and prephenic acid **5.3**.

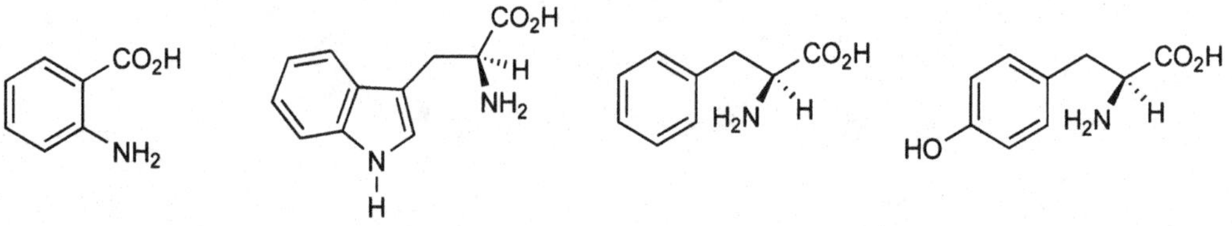

Figure 5.2   Amino acids produced from the shikimic acid pathway.

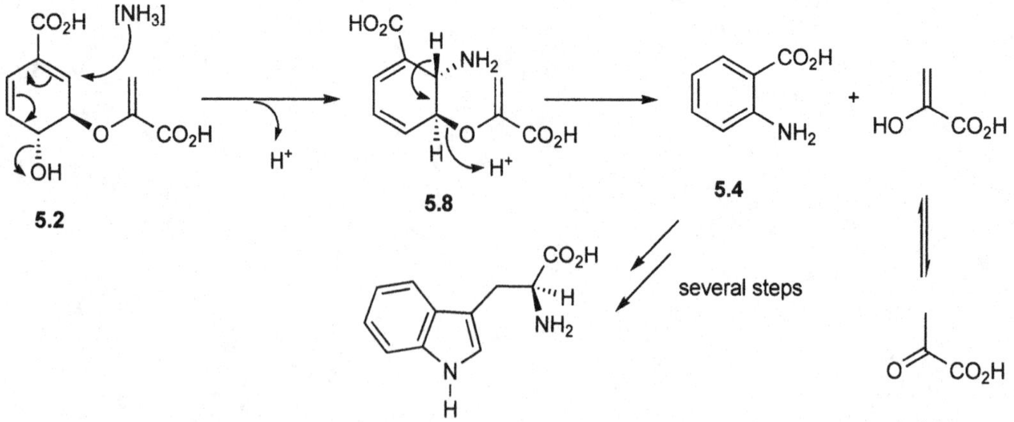

Figure 5.3   Biosynthesis of anthranilic acid **5.4**.

ring of shikimic acid **5.1**. These four compounds **5.4–5.7** are the principal biosynthetic intermediates in the shikimic acid pathway.

Anthranilic acid **5.4** is biosynthesized from chorismic acid **5.2** as shown in Figure 5.3. Reaction of chorismic acid with a source of ammonia (this is provided by the amino acid, glutamic acid) gives the intermediate **5.8**, which subsequently undergoes elimination of the enol tautomer of pyruvic acid **5.9** thus yielding anthranilic acid **5.4**. Anthranilic acid is then converted through an elaborate sequence of reactions into tryptophan **5.5**.

Problem

Problem 5.1 Suggest how chorismic acid **5.2** might react with water to produce salicyclic acid (2-hydroxybenzoic acid).

5.2 Transamination

Before we consider the biosynthesis of phenylalanine **5.6** and tyrosine **5.7** from prephenic acid **5.3**, we briefly need to mention a common biochemical reaction called *transamination*. Transamination is controlled by enzymes termed *transaminases*, and the overall process is shown in Figure 5.4 with the side chains of the substrate and product molecules represented by the circles and the squares. The substrates for a transamination reaction are an *amino acid* and an *α-keto acid* (carboxylic acid derivatives in which the carboxylic acid group is directly attached to a carbonyl group are termed *α*-keto acids). The amino acid substrate is converted into the *α*-keto acid product and concomitantly an *α*-keto acid is converted into an amino acid. Thus, you can consider transamination as an 'exchange' of an amino group for a ketone group and vice versa.

The mechanism of the transamination reaction is shown in Figure 5.5. The amino group of an amino acid reacts with the aldehyde group of the transaminase enzyme co-factor, pyridoxal phosphate **5.10** to give an imine **5.11**. Tautomerization of imine **5.11** gives the new imine **5.12** which is then hydrolysed giving the *α*-keto acid and pyridoxamine phosphate

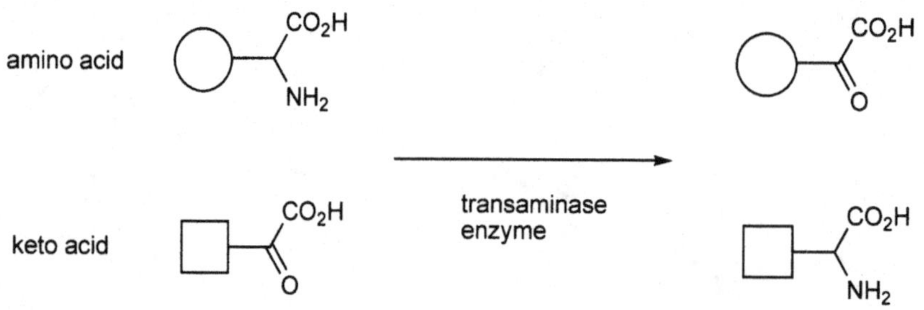

Figure 5.4   Transamination.

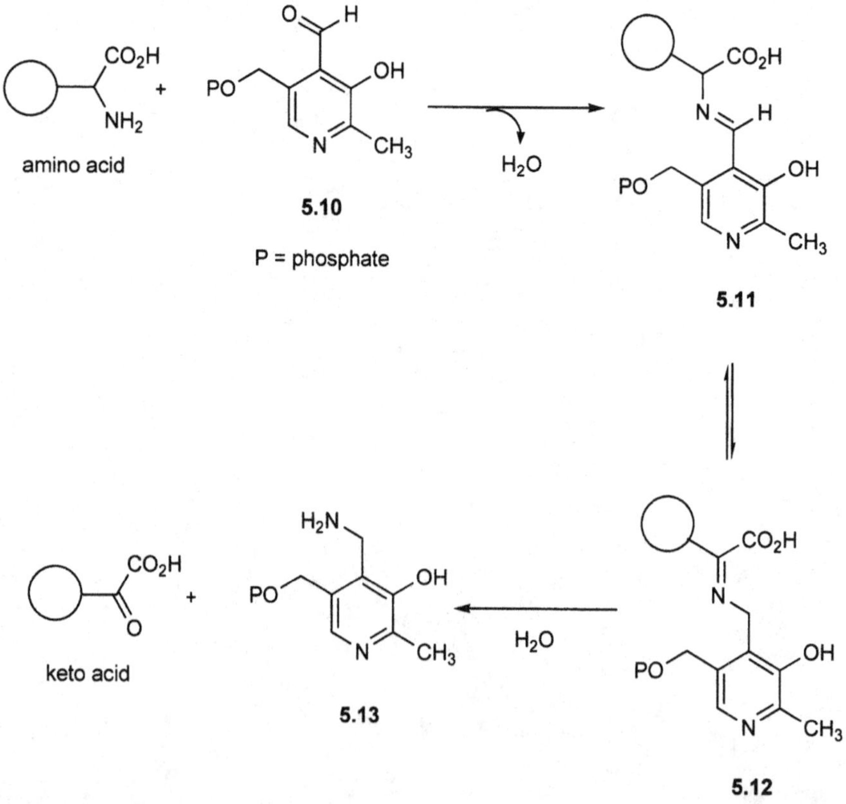

Figure 5.5   Mechanism of transamination.

70

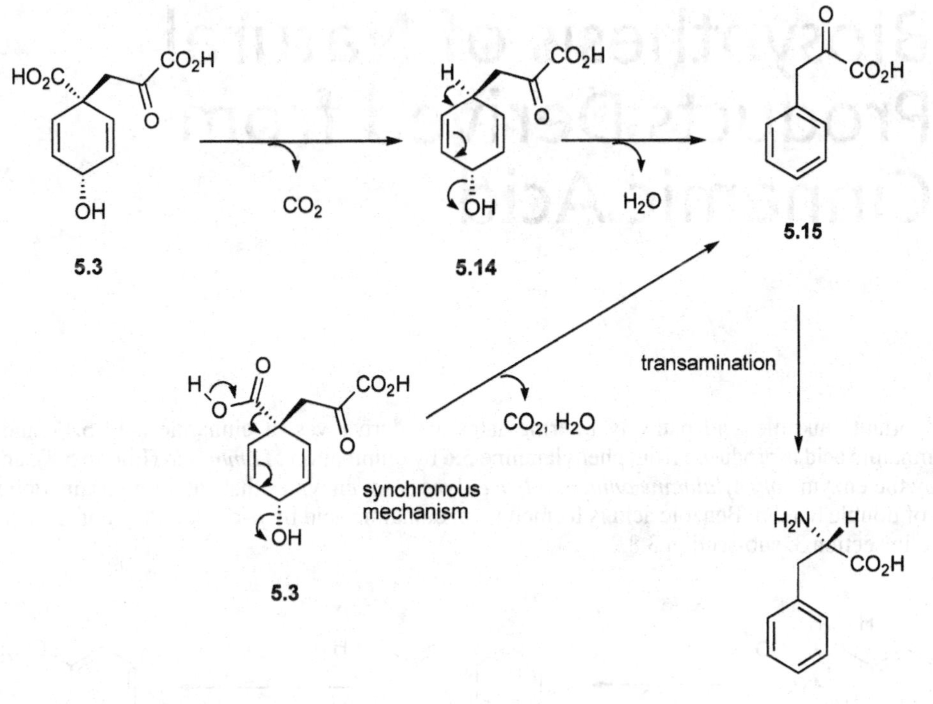

Figure 5.6 Biosynthesis of phenylalanine **5.6**.

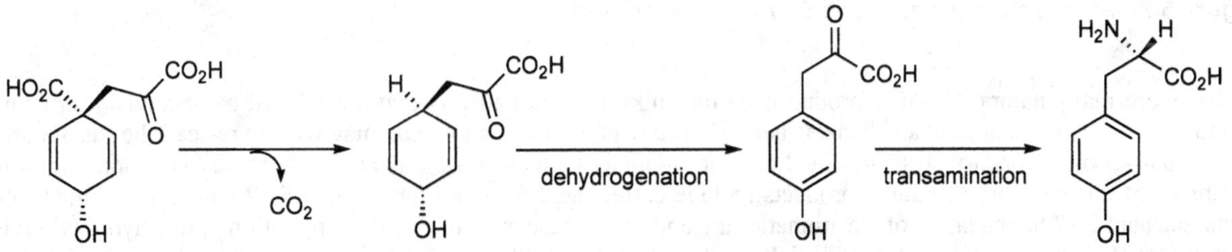

Figure 5.7 Biosynthesis of tyrosine **5.7**.

5.13. Since transamination can proceed in either direction, pyridoxamine phosphate and an α-keto acid would produce an amino acid and thus regenerate the pyridoxal phosphate **5.10**.

The biosynthesis of phenylalanine **5.6** from prephenic acid **5.3** is shown in Figure 5.6 Prephenic acid is decarboxylated giving the intermediate **5.14** which then eliminates water giving phenylpyruvic acid **5.15**. Transamination of phenylpyruvic acid as discussed above then yields phenylalanine **5.6**. Although the transformation of prephenic acid to phenylpyruvic acid has been shown as a stepwise process, i.e. carbon dioxide and water are lost in separate reactions, a synchronous loss of these two molecules probably occurs and this is also shown in Figure 5.6.

The biosynthesis of tyrosine **5.7** from prephenic acid **5.3** is shown in Figure 5.7. Loss of carbon dioxide from prephenic acid gives the intermediate **5.14**, and subsequent oxidation of this intermediate then yields the hydroxyphenylpyruvic acid **5.16**. This oxidation step is an example of *dehydrogenation*, so called because two hydrogen atoms are being removed. Dehydrogenation is a common type of organic reaction and occurs readily if there is a suitable driving force – in this case an aromatic benzene ring is being formed. Finally, transamination of compound **5.16** then affords tyrosine **5.7**.

5.3 Biosynthesis of Natural Products Derived from Cinnamic Acid

Two other important shikimic acid pathway intermediates are derivatives of cinnamic acid **5.17** and benzoic acid ($PhCO_2H$). Cinnamic acid is produced from phenylalanine **5.6** by elimination of *ammonia* (Figure 5.8), and this reaction is controlled by the enzyme *phenylalanine ammonia lyase* (lyases are enzymes that catalyse elimination reactions with the formation of double bonds). Benzoic acid is formed from cinnamic acid by β-oxidation – a process that has already been discussed in section 3, sub-section 3.8.

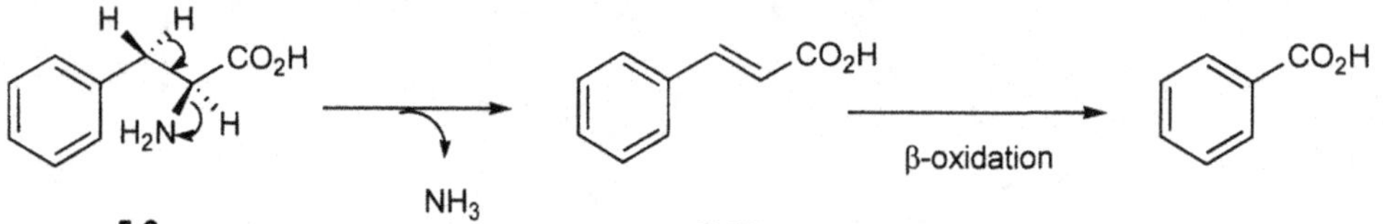

Figure 5.8 Formation of cinnamic acid **5.17** and benzoic acid.

There are many natural products produced by the shikimic acid pathway that are formed by secondary structural modifications to cinnamic acid and benzoic acid. Before reading this section, you may wish to re-read the sub-sections in section 4.5 on secondary structural modifications – many of these reactions are appropriate here. Some relevant examples of this category of natural products include caffeic acid **5.18** and ferulic acid **5.19** which are formed from cinnamic acid **5.17** by oxidation of the aromatic ring and, in the case of ferulic acid, methylation of one hydroxyl group also occurs (Figure 5.9). Avenanthramide **5.20**, a phytoalexin of oat leaves, is an amide derivative that is formed from cinnamic acid (as its coenzyme A thioester derivative) and anthranilic acid **5.4** (Figure 5.9). Both aromatic rings in compound **5.20** are hydroxylated at some stage in its biosynthesis.

Problem

Problem 5.2 Vanillic acid is produced from ferulic acid **5.19** by β-oxidation. Propose a reasonable structure for vanillic acid.

When hydroxylation of cinnamic acid **5.17** occurs at the *ortho*-position, intramolecular cyclization of the hydroxyl group with the *cis* acid group can occur giving derivatives of coumarin **5.21**, as depicted in Figure 5.10. The coumarin class of natural products occurs widely, especially in the *Umbelliferae* family of plants, and examples include umbelliferone **5.22** and ascletin **5.23** (Figure 5.10).

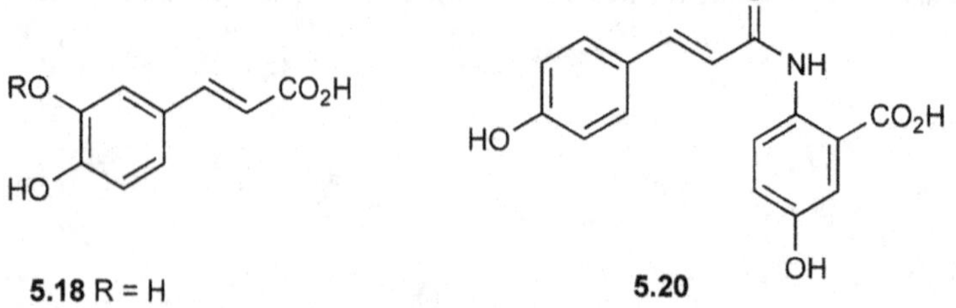

5.18 R = H
5.19 R = CH_3

5.20

Figure 5.9 Caffeic acid **5.18**, ferulic acid **5.19** and avenanthramide **5.20**.

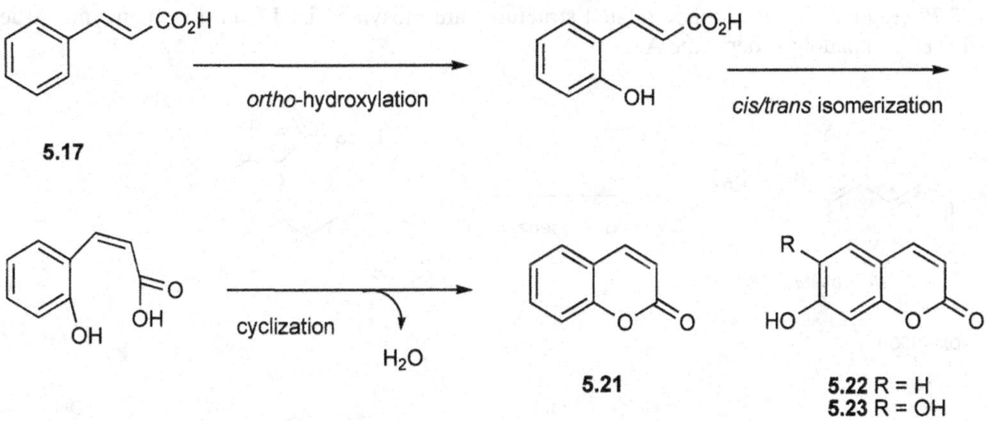

Figure 5.10 Umbelliferone **5.22** and ascletin **5.23**.

Problem

Problem 5.3 Isoprenylation of umbelliferone **5.22** gives compound **5.24** (see section 4, sub-section 4.5.1). Show how compound **5.24** might be transformed into both xanthyletin **5.25** and psoralen **5.26**.

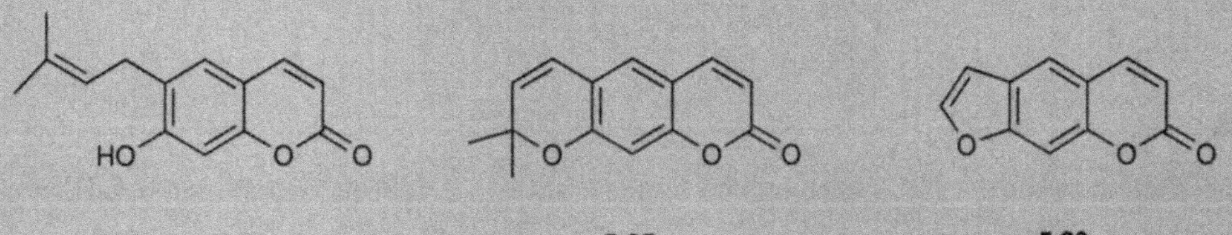

Numerous types of natural products exist that are of mixed biosynthetic origin; part of the structure is derived from the shikimic acid pathway and part from the polyketide pathway. Gingerol **5.28**, the spice obtained from the ginger plant, is one such compound which is formed from the acetyl coenzyme A derivative of ferulic acid **5.19** and the thioester **5.27**, as outlined in Figure 5.11.

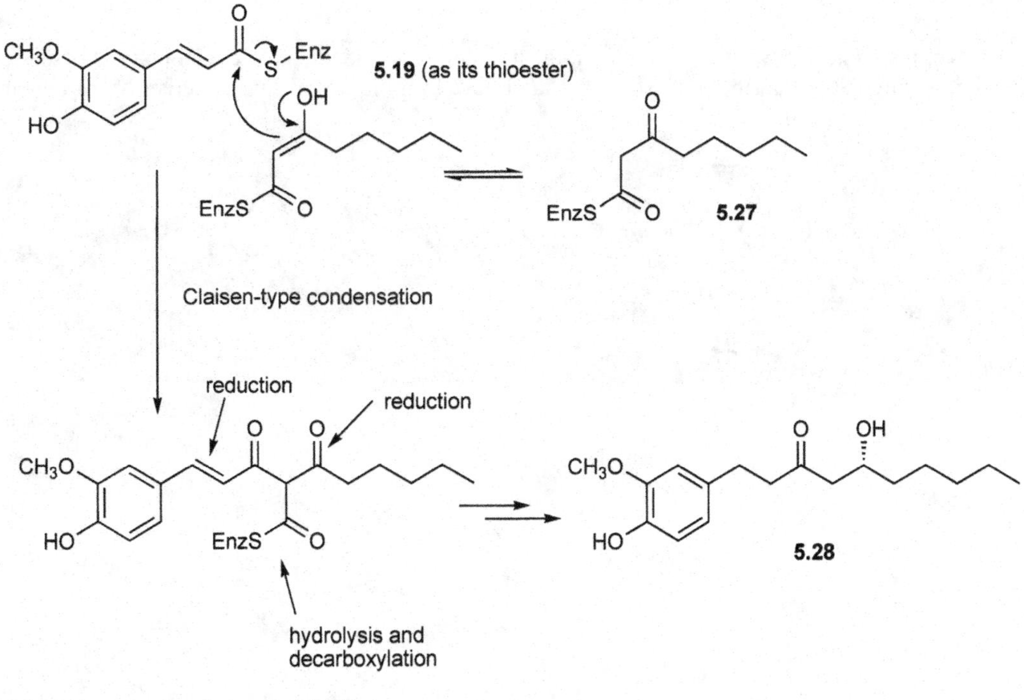

Figure 5.11 Biosynthesis of gingerol **5.28**.

Reveratrol **5.29** (Figure 5.12) and many related structures are biosynthesized from the coenzyme A derivative of cinnamic acid **5.17** and malonyl coenzyme A.

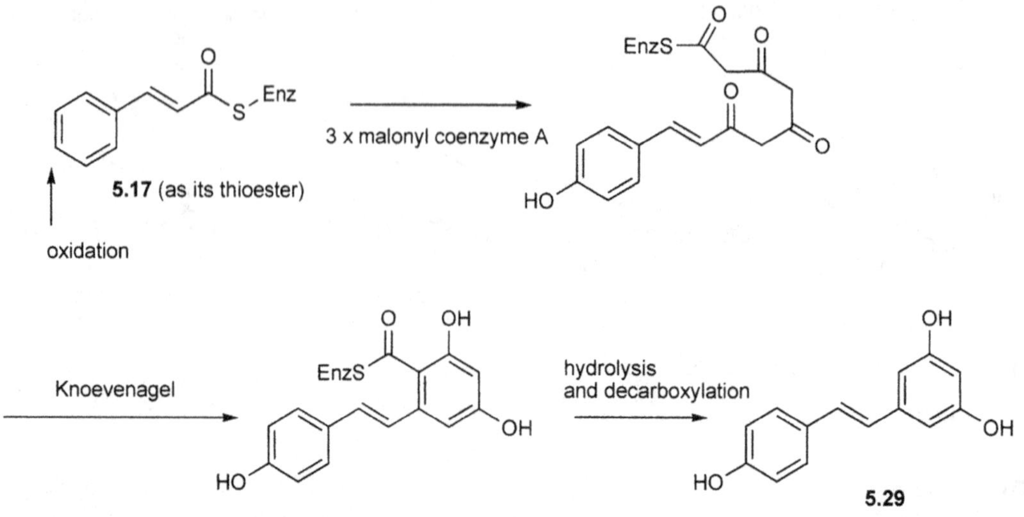

Figure 5.12 Biosynthesis of reveratrol **5.29** and related compounds.

Problems

Problem 5.4 Outline reasonable biosynthetic routes to morachalcone A **5.30** and stemofuran C **5.31** from the coenzyme A derivative of cinnamic acid **5.17**.

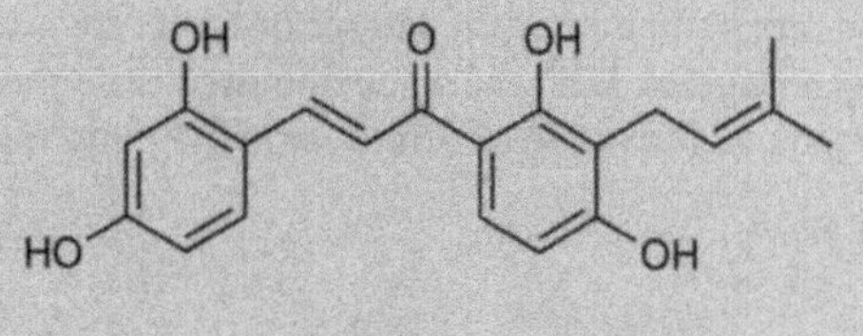

5.30 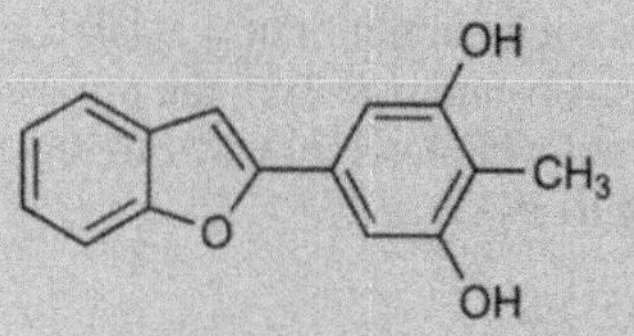   **5.31**

Problem 5.5 Stemanthrene C **5.33** is formed by a radical coupling of the phenolic compound **5.32** (see section 4, sub-section 4.5.3). Suggest a plausible route to compound **5.32** from the coenzyme A derivative of cinnamic acid **5.17**.

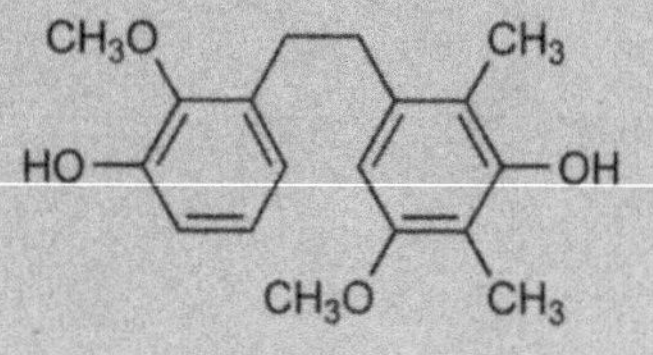

5.32 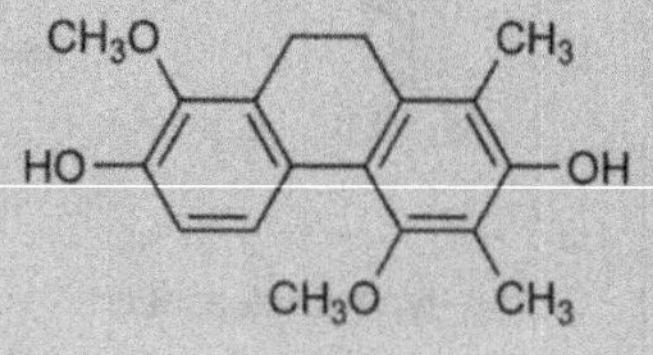   **5.33**

Problem 5.6 Suggest a reasonable biosynthetic pathway to yangonin **5.34** from the coenzyme A derivative of cinnamic acid **5.17**.

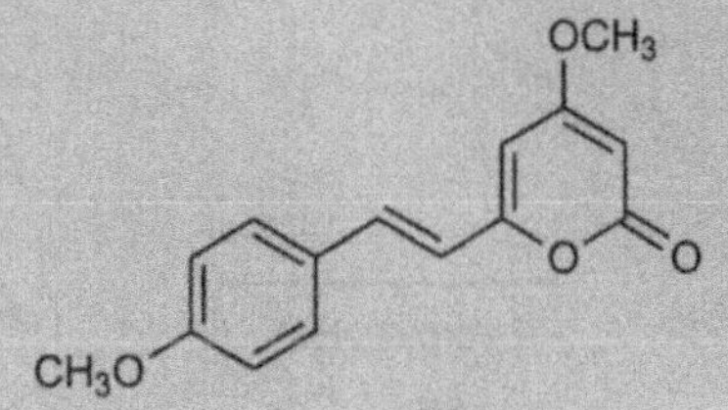

5.34

Flavones **5.35** and isoflavones **5.36** have general structures as shown in Figure 5.13 and these natural products are widely distributed in plants. Flavones, as exemplified by apigenin **5.38** (Figure 5.14) are produced from the coenzyme A derivative of cinnamic acid **5.17** and malonyl coenzyme A via the flavanone intermediate, naringenin **5.37**. Flavanones are dihydro-flavones.

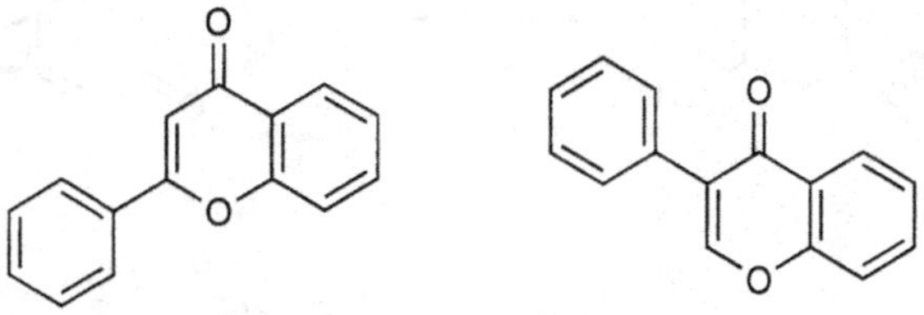

5.35   **5.36**

Figure 5.13 Flavones **5.35** and isoflavones **5.36**.

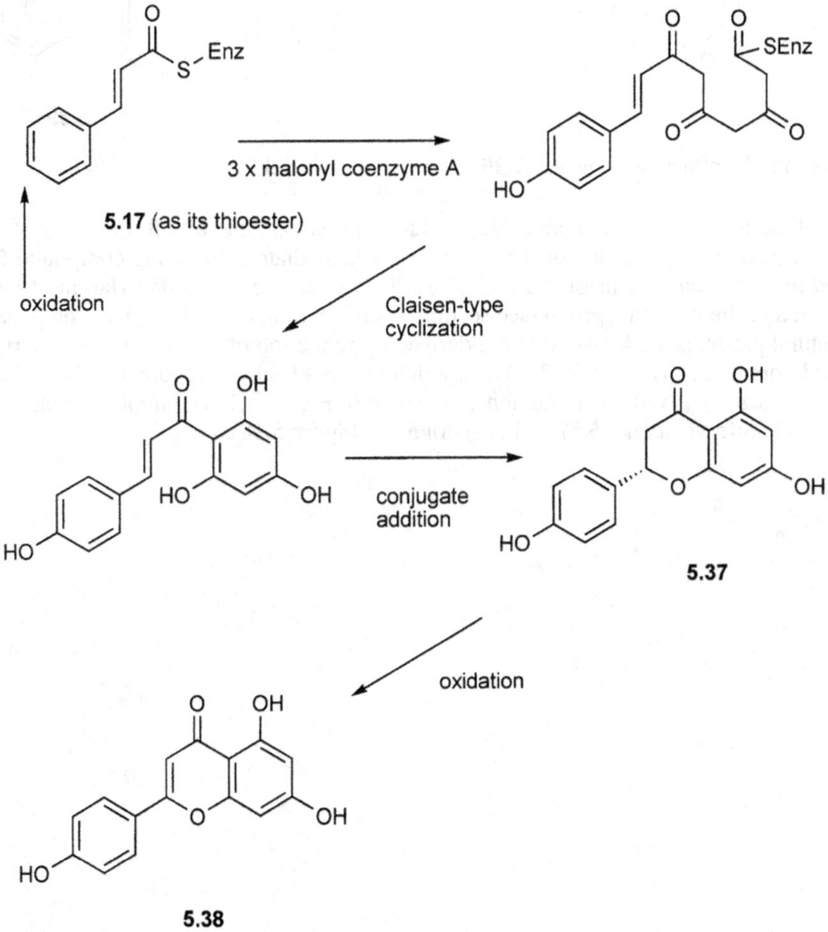

Figure 5.14 Biosynthesis of apigenin **5.38**.

Isoflavones are produced from flavanones by migration of the aromatic group, as illustrated in Figure 5.15 for the biosynthesis of genistein **5.39** from naringenin **5.37**. The migration of the aromatic group may involve initial hydroxylation of the aromatic ring as illustrated in Figure 5.15.

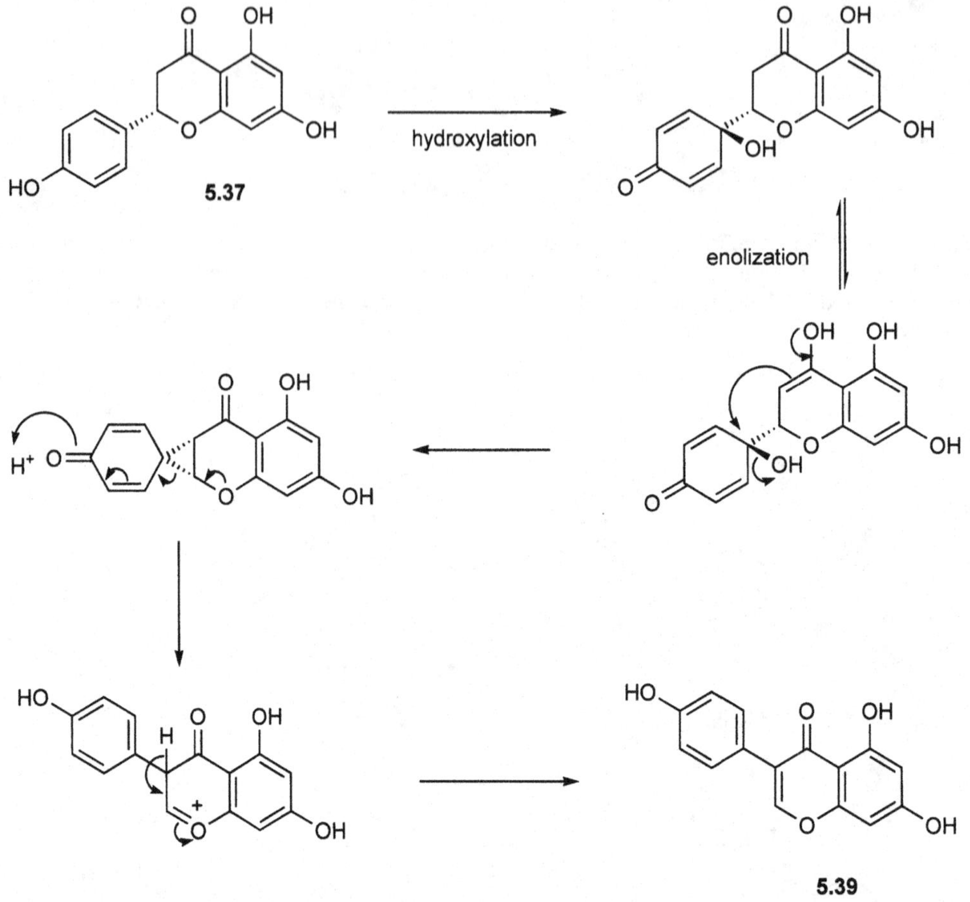

Figure 5.15 Formation of isoflavone genistein **5.39** from naringenin **5.37**.

The biosynthesis of the insecticide, rotanone **5.42**, provides a good example of a flavanone–isoflavone interconversion (Figure 5.16). An oxidative cyclization of the isoflavone intermediate **5.40** giving compound **5.41** occurs during the biosynthesis and this reaction may involve a radical cyclization as shown. Isoprenylation of compound **5.41** and subsequent modification of the isoprenyl group (see section 4, sub-section 4.5.1) then gives rotanone **5.42**.

Many types of natural products are known that are derived by reduction of the coenzyme A derivative of cinnamic acid **5.17** and related compounds (Figure 5.17). Thus, aldehydes **5.43**, allylic alcohols **5.44** and allylic compounds **5.45** can all be biosynthesized by NADPH reductions, as shown in Figure 5.17. Examples of natural products include cinnamaldehyde **5.56**, conniferyl alcohol **5.57** and eugenol **5.58** (Figure 5.18).

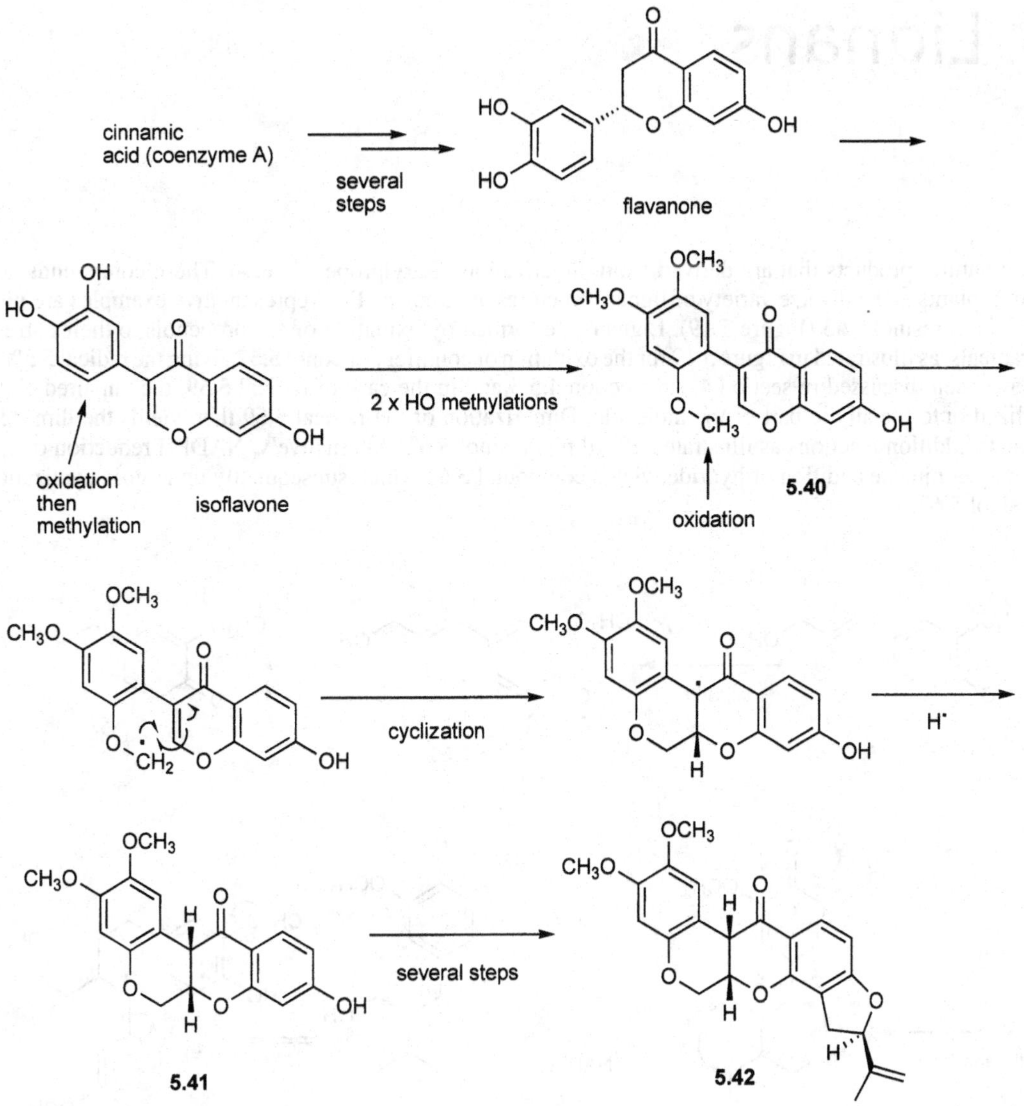

Figure 5.16 Biosynthesis of rotanone **5.42**.

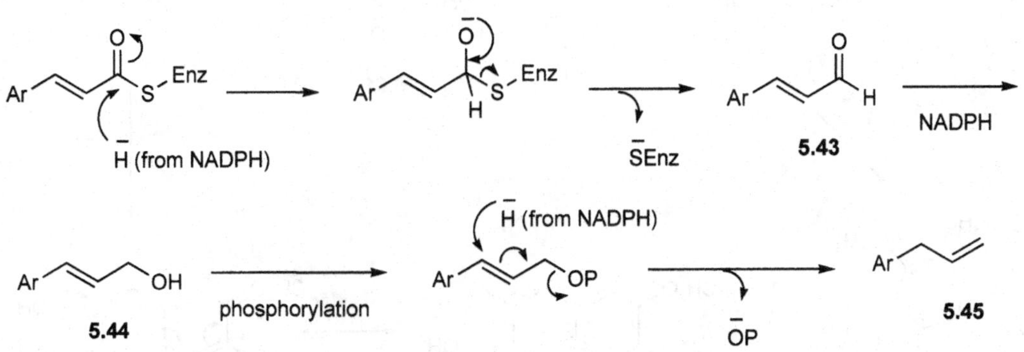

Figure 5.17 Compounds derived from reduction of cinnamic acid derivatives.

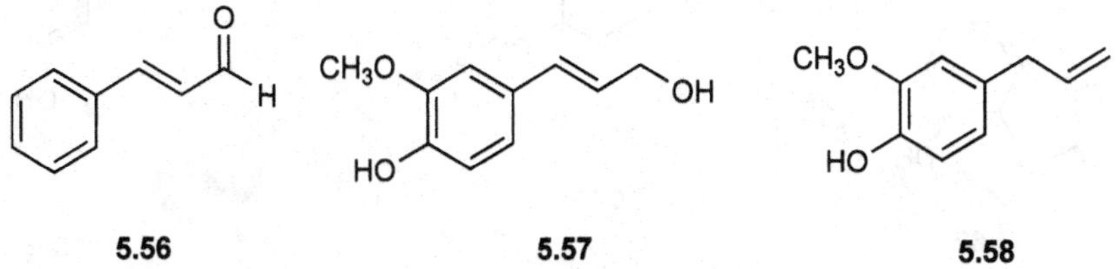

Figure 5.18 Cinnamaldehyde **5.56**, conniferyl alcohol **5.57** and eugenol **5.58**.

5.4 Lignans

Lignans are natural products that are derived from dimerization of arylpropenols **5.44**. These compounds are widely distributed in plants and a diverse variety of lignan structures are known. Two representative examples are pinoresinol **5.62** and isolariciresinol **5.63** (Figure 5.19). Lignans are formed by oxidation of arylpropenols to their corresponding phenoxy radicals, as illustrated in Figure 5.19 for the oxidation of conniferyl alcohol **5.57** giving the radical **5.59**. Phenoxy radicals have been discussed in section 4, sub-section 4.5.3 and in the case of radical **5.59**, the unpaired electron can be delocalized onto the allylic part of the molecule. Dimerization of the radical **5.59** then yields the dimer **5.60**, and two conjugate addition reactions as illustrated afford pinoresinol **5.62**. Alternatively, NADPH reduction of compound **5.60**, again by conjugate addition of hydride, yields compound **5.61** which subsequently undergoes cyclization giving isolariciresinol **5.63**.

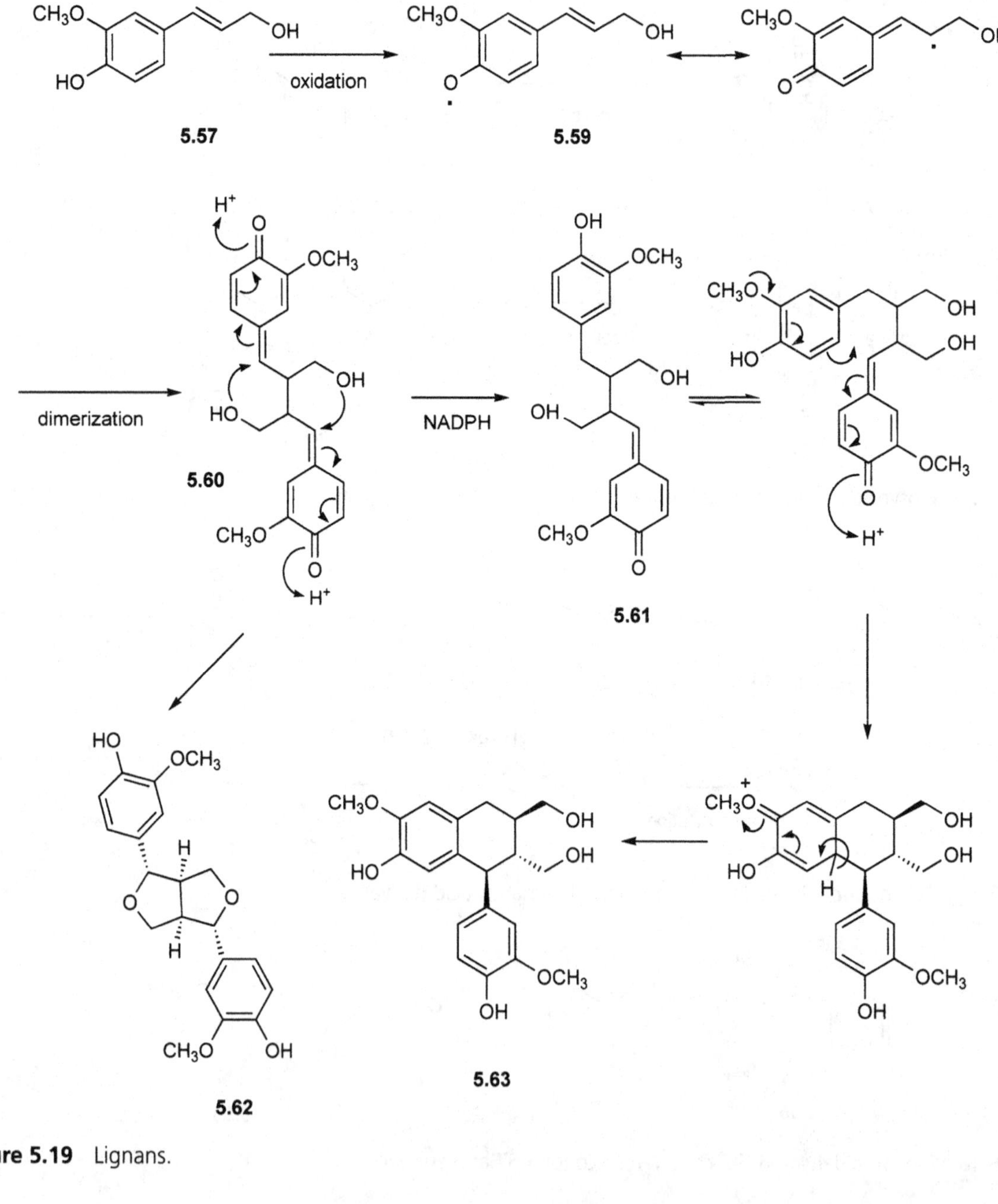

Figure 5.19 Lignans.

Problems

Problem 5.7 Propose reasonable biosynthetic routes to the lignans phyllanthin **5.64** and arctigenin **5.65** from conniferyl alcohol **5.57**.

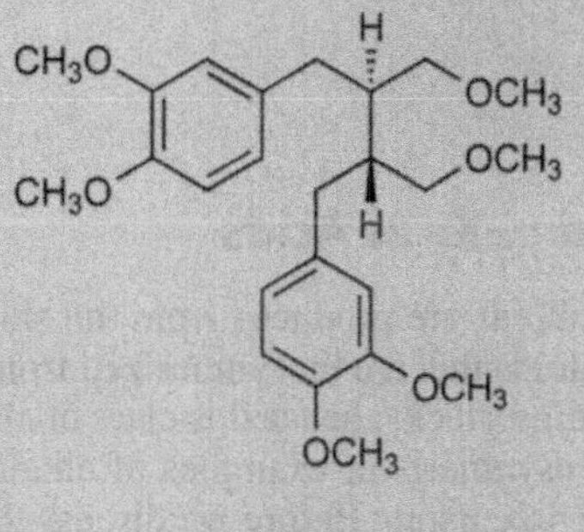

5.64

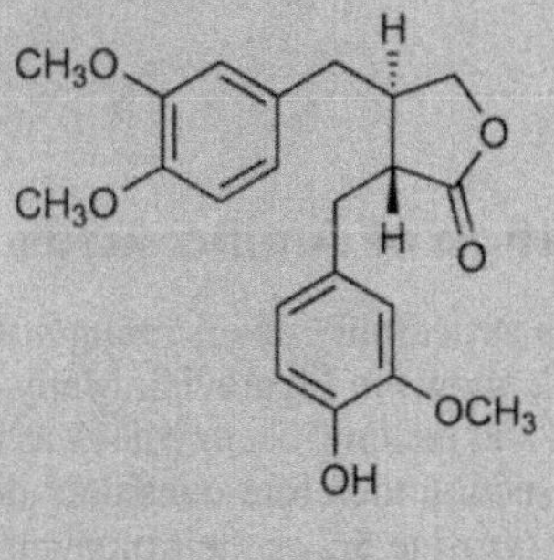

5.65

Problem 5.8 Isoeugenol is an isomer of eugenol **5.58**. Suggest a structure for this compound and deduce how it might be formed by phosphorylation and NADPH reduction of compound **5.57**.

Problem 5.9 Show how the lignans dihydroguaiaretic acid **5.67** and veraguensin **5.68** could be formed from isoeugenol.

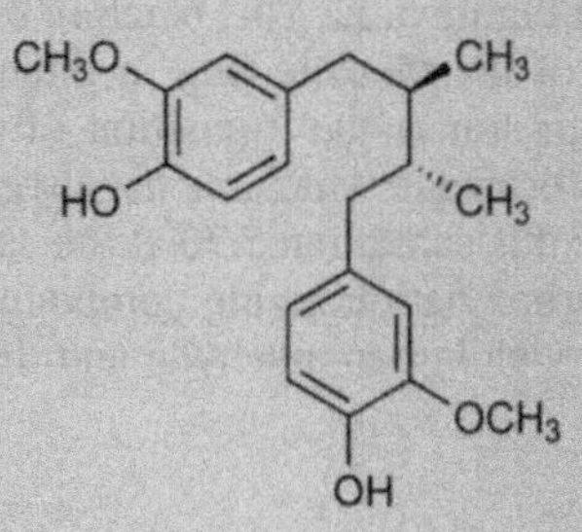

5.67

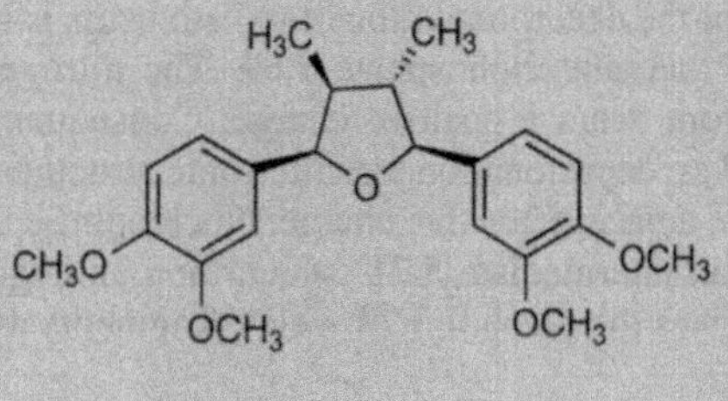

5.68

Many shikimic acid derived natural products possess a methylenedioxy fragment (—OCH$_2$O—) as illustrated in structure **5.66**. This fragment is formed presumably by an oxidation process as illustrated in Figure 5.20. Thus, many lignans possess this substituent because the *ortho*-methoxyphenol functionality present in conniferyl alcohol **5.57** can be converted into the methylenedioxy group. Similarly, many alkaloids contain methylenedioxy groups as we shall see later in this section.

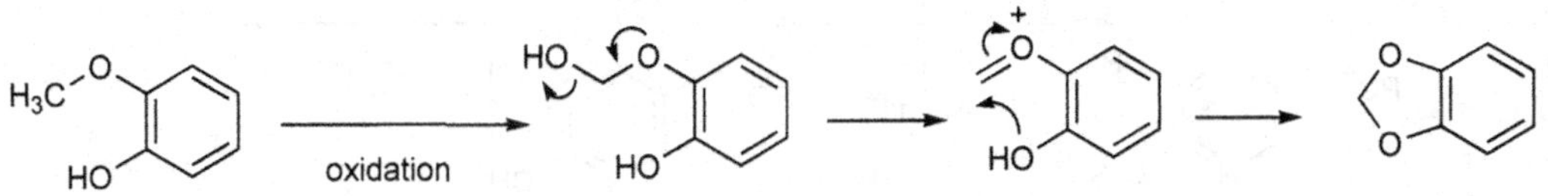

5.66

Figure 5.20 Formation of methylenedioxy groups.

5.5 Biosynthesis of Alkaloids

5.5.1 Decarboxylation of Amino Acids and α-Keto Acids

An immense number and a structurally diverse range of alkaloids are produced from the shikimic acid pathway from the building blocks **5.4–5.7**, shown in Figure 5.2. Many alkaloids that are biosynthesized from anthranilic acid **5.4** also require acetyl coenzyme A and malonyl coenzyme A as building blocks; hence this class of alkaloid has been discussed as part of section 4. This section therefore considers the biosynthesis of examples of alkaloids that have tryptophan **5.5**, phenylalanine **5.6** and tyrosine **5.7** as their biosynthetic precursors. Before we discuss these types of alkaloid, we briefly need to consider how the decarboxylation reactions of the amino acid and α-keto acids that we met in Figure 5.4 can occur, i.e. $RCOCO_2H \rightarrow RCHO$ and $RCH(NH_2)CO_2H \rightarrow RCH_2NH_2$ respectively. These decarboxylated products (i.e. aldehydes and amines) are also important building blocks in alkaloid biosynthesis. The decarboxylation of amino acids requires the cofactor pyridoxal phosphate **5.10** which reacts with an amino acid giving the intermediate **5.11** (see Figure 5.5). The intermediate **5.11** can then undergo decarboxylation as depicted in Figure 5.21 giving a carbanion in which the negative charge can be delocalized. Protonation of the carbanion and hydrolysis of the resulting imine gives an amine and pyridoxal phosphate **5.10**.

The mechanism for the decarboxylation of α-keto acids is shown in Figure 5.22. This reaction requires the thiazole-containing cofactor, thiamine diphosphate **5.69**. The nitrogen atom of the thiazole ring in thiamine diphosphate is alkylated and therefore bears a positive charge. Consequently, the proton at the 2-position of the thiazole ring is relatively acidic, and on deprotonation a zwitterionic structure **5.70** is formed in which the negative charge is stabilized to some extent by the adjacent positive charge. Nucleophilic attack by this carbanion **5.70** at the carbonyl group of the α-keto acid affords an intermediate **5.71**, which then undergoes decarboxylation giving compound **5.72**. Protonation of compound **5.72** gives the alcohol **5.73** which fragments as shown yielding an aldehyde and thiamine diphosphate **5.69**.

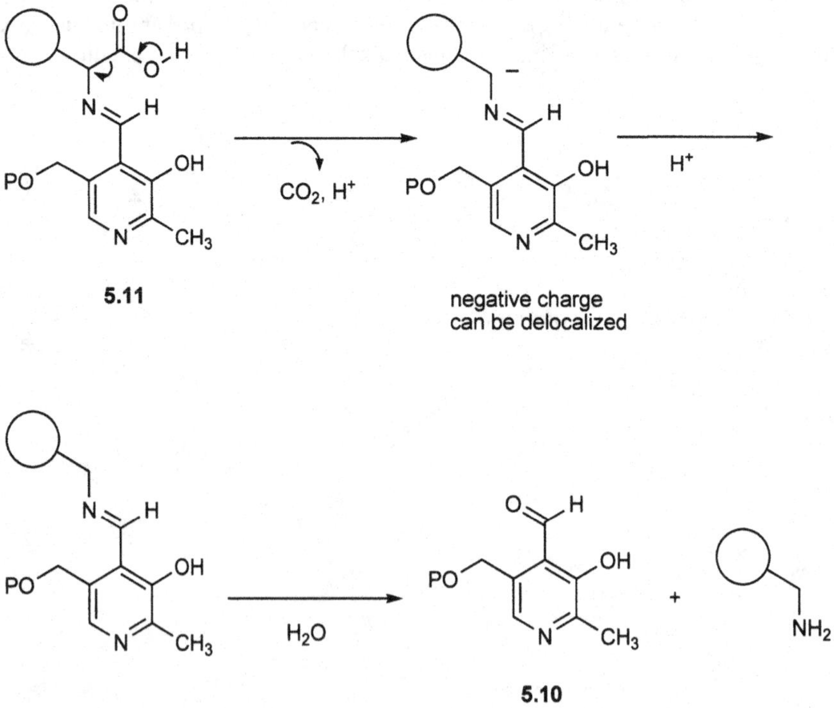

Figure 5.21 Decarboxylation of amino acids.

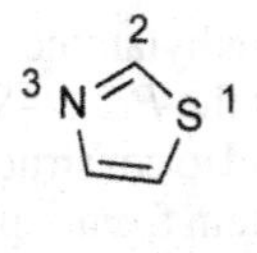

thiazole

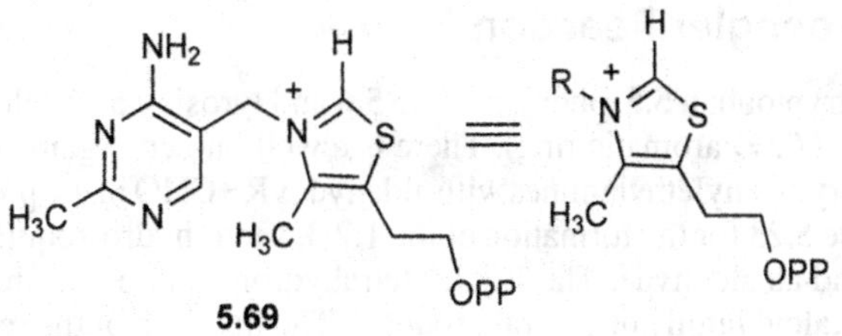

5.69

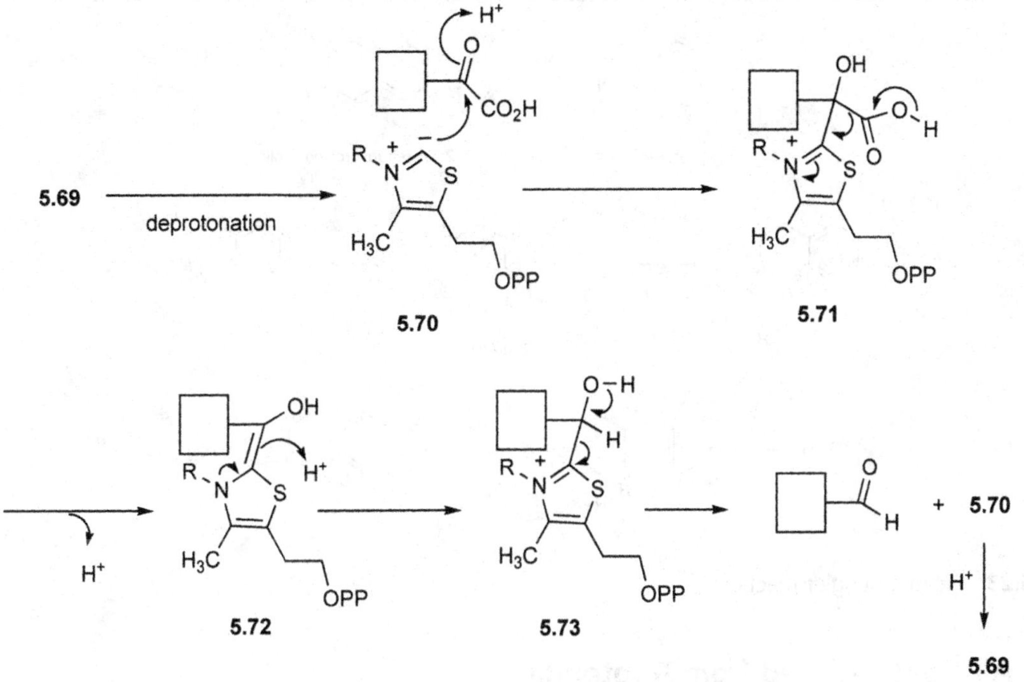

Figure 5.22 Decarboxylation of α-keto acids.

Problem

Problem 5.10 Suggest how the neurotransmitter serotonin **5.74** might be biosynthesized from tryptophan **5.5**, and also how adrenaline **5.75** and mescaline **5.76** might be formed from tyrosine **5.7**.

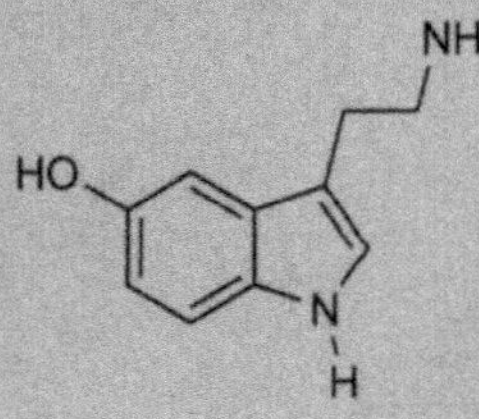

5.74

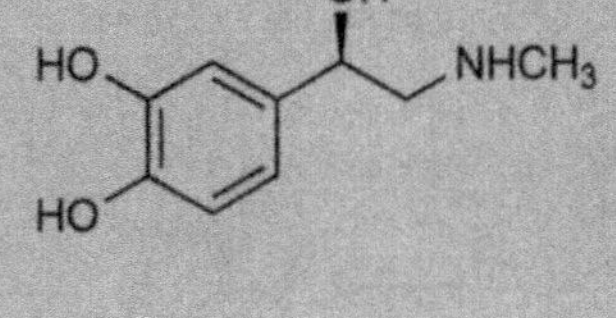

5.75

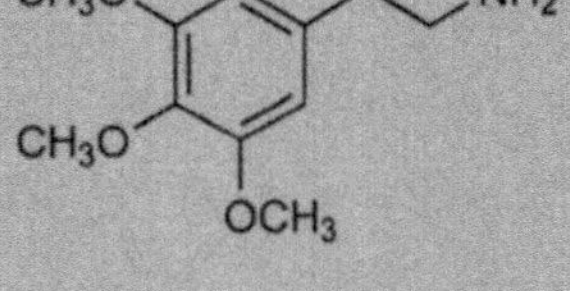

5.76

5.5.2 Pictet–Spengler Reaction

Decarboxylation of tryptophan **5.5**, phenylalanine **5.6** and tyrosine **5.7** yields 2-arylphenylethylamines of general structure ArCH$_2$CH$_2$NH$_2$ (Ar = aromatic ring). There is a well known organic reaction called the *Pictet–Spengler* reaction which condenses 2-arylphenylethylamines with aldehydes R—CHO in the presence of an acid to construct a new ring. This is illustrated in Figure 5.23 for the formation of the 1,2,3,4-tetrahydroisoquinoline ring system from 2-phenylethylamine (PhCH$_2$CH$_2$NH$_2$) and an aldehyde. The 1,2,3,4-tetrahydroisoquinoline ring system and related heterocycles are found extensively in the alkaloid family of natural products. The first step of the reaction is the condensation of the amine with the aldehyde to form an imine **5.77**. The protonated imine is electron-deficient and subsequently acts as an electrophile in an intramolecular electrophilic substitution reaction, as shown in Figure 5.23, giving the 1,2,3,4-tetrahydroisoquinoline derivative **5.78**. This reaction is mechanistically similar to typical electrophilic substitution reactions of benzene.

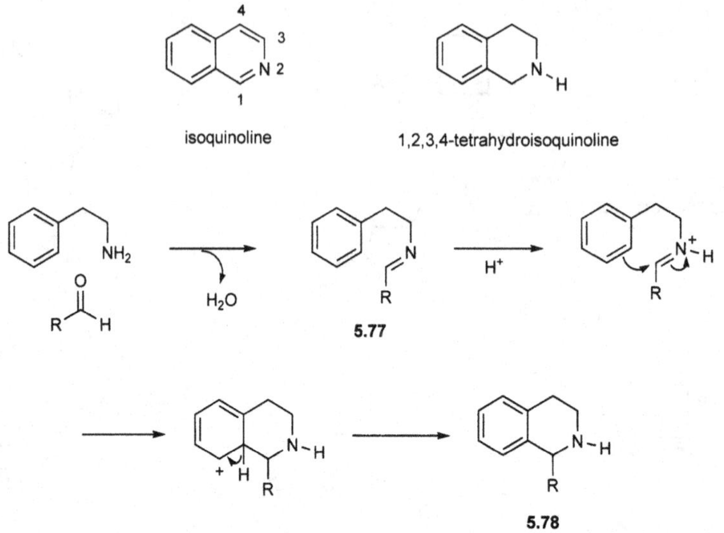

Figure 5.23 Pictet–Spengler reaction.

5.5.3 Alkaloids Derived from Tryptophan

A representative example of an indole-ring containing alkaloid that is derived from tryptophan **5.5** via decarboxylation and Pictet–Spengler reactions is harmane **5.80** (Figure 5.24). The biosynthetic route to harmane is relatively straightforward: decarboxylation of tryptophan **5.5** and a subsequent Pictet–Spengler reaction with acetaldehyde (CH$_3$CHO) gives the tetrahydro derivative **5.79**, from which harmane **5.80** is formed by oxidation.

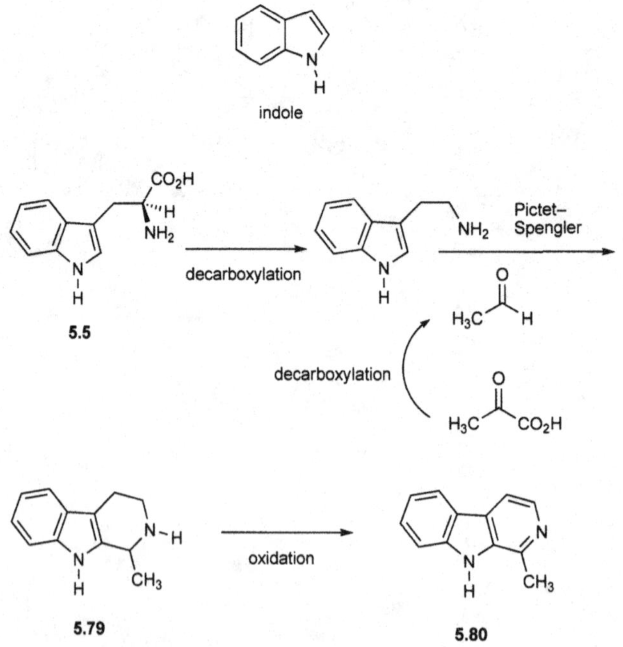

Figure 5.24 Biosynthesis of harmane **5.80**.

Ajmalicine **5.84** is an example of a more structurally complex indole alkaloid which is also formed from tryptophan **5.5** via decarboxylation and Pictet–Spengler reactions (Figure 5.25). The aldehyde component in the Pictet–Spengler reaction is secologanin **5.81** which is a terpenoid natural product (section 6, sub-section 6.4.1). Ajmalicine, like so many other natural products is therefore a compound derived by a mixed metabolic pathway. In secologanin **5.81**, Glu represents a glucose unit that is removed by hydrolysis after the Pictet–Spengler reaction giving the intermediate **5.82**. Ring-opening of the oxygen-containing ring then occurs liberating an aldehyde group in structure **5.83**. This aldehyde is then transformed into ajmalicine **5.84** by the sequence of reactions shown in Figure 5.25. The combination of tryptophan **5.5** and secologanin gives rise to many structurally complex natural products, including the well known compounds strychnine and quinine.

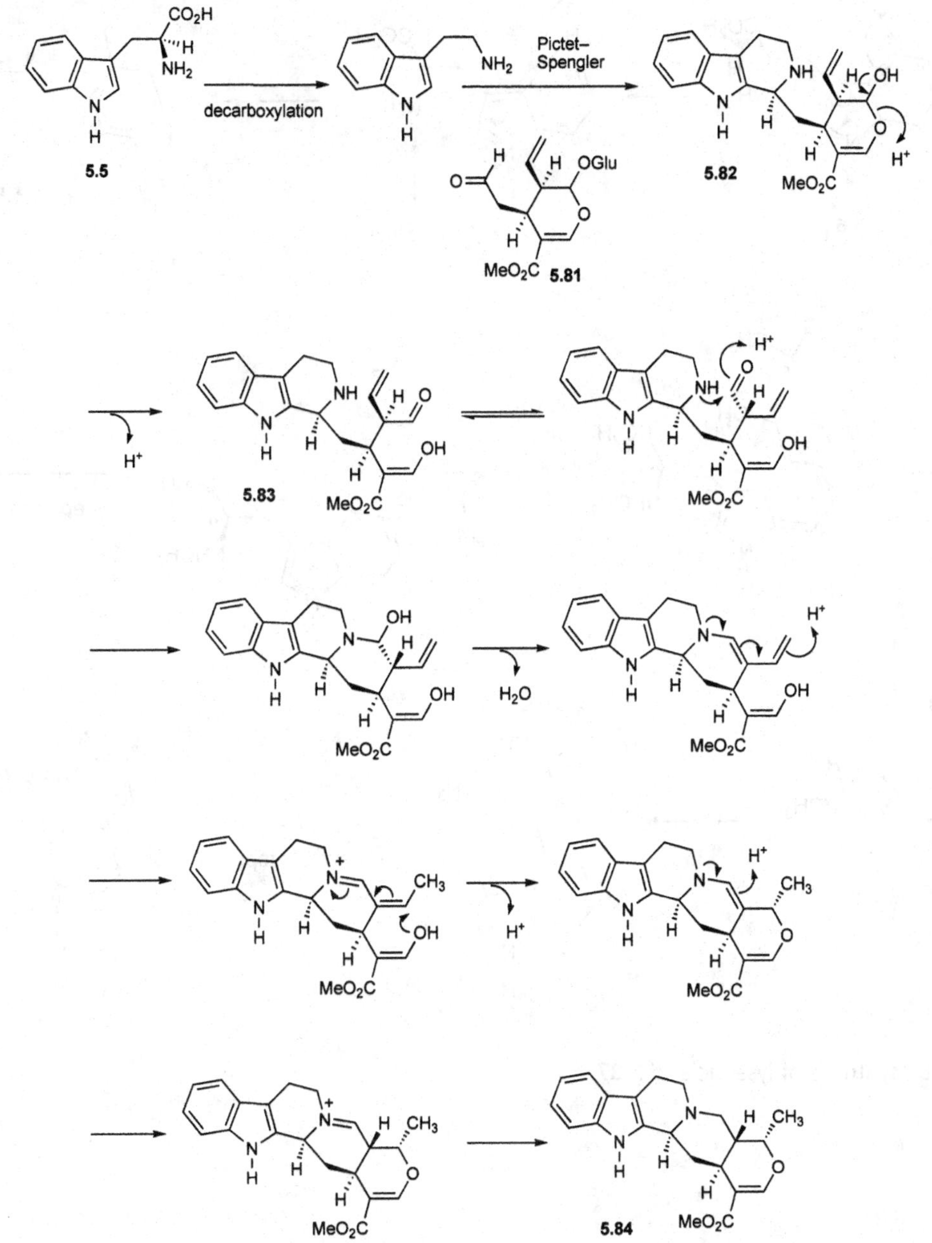

Figure 5.25 Biosynthesis of ajmalicine **5.84**.

The indole ring is electron-rich and consequently it can react with biological alkylating agents. This property of indole has been exploited in Nature with the biosynthesis of the ergot class of alkaloids of which lysergic acid **5.87** is a well known example (Figure 5.26). Isoprenylation of tryptophan **5.5** with the terpene building-block dimethylallyl pyrophosphate **5.85** gives the intermediate **5.86** from which lysergic acid is formed, as outlined in Figure 5.26. Note the arrangement of the numbered atoms in the penultimate structure and lysergic acid **5.87** which has been established by isotopic labelling.

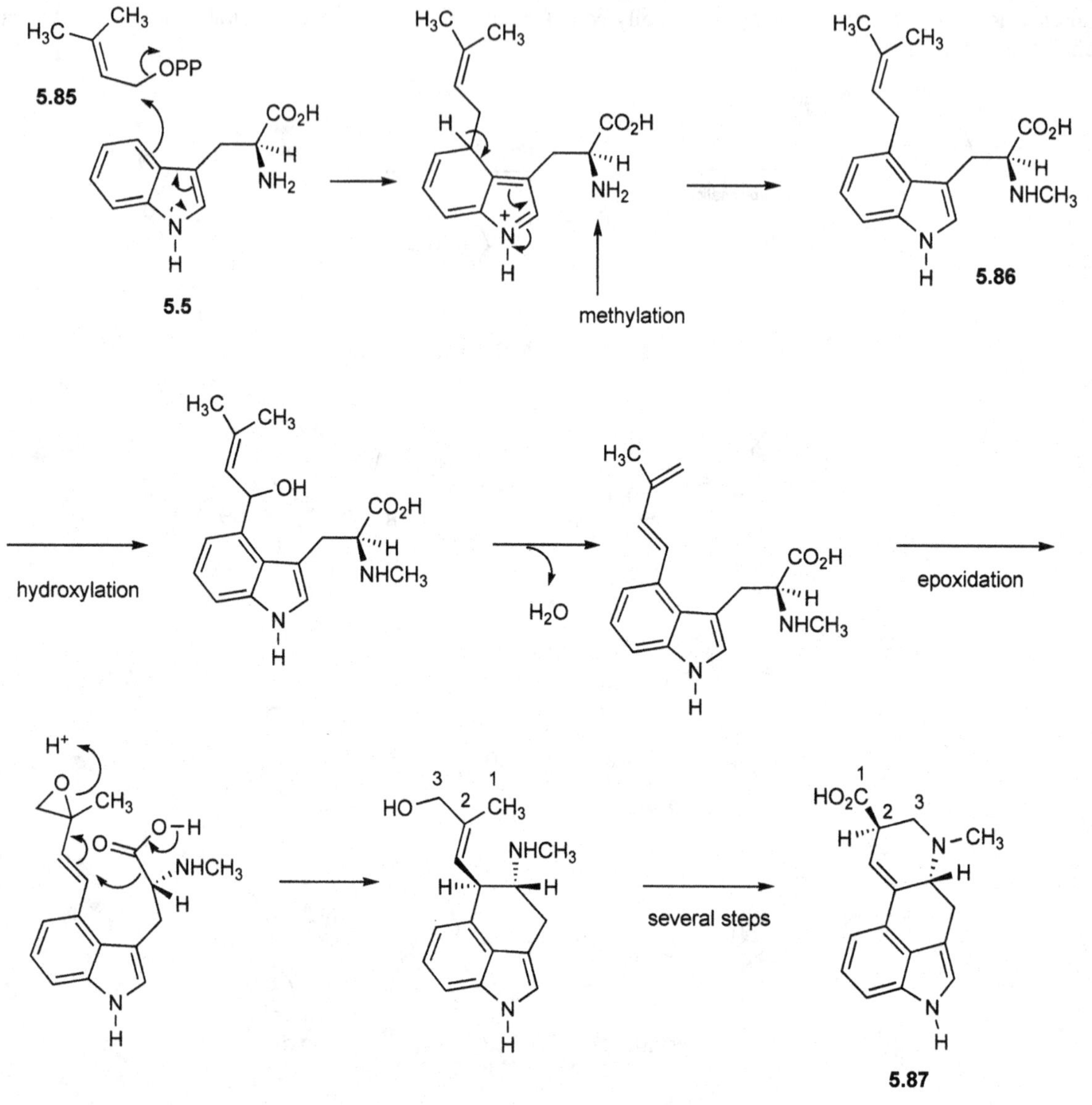

Figure 5.26 Biosynthesis of lysergic acid **5.87**.

The biosynthesis of lysergic acid illustrates alkylation (isoprenylation) of the indole skeleton at the 4-position. Numerous alkaloids are also produced as a result of alkylation of the indole nucleus at the 3-position, as illustrated for the biosynthesis of the alkaloid **5.89** from tryptophan **5.5** (Figure 5.27). Thus, isoprenylation at the 3-position of the indole ring occurs (you can view the pyrrole part of the ring as an enamine) giving an intermediate iminium ion **5.88** which then cyclizes forming the new nitrogen-containing ring. Subsequently, isoprenylation and methylation occur at the nitrogen atoms, giving the alkaloid **5.89**.

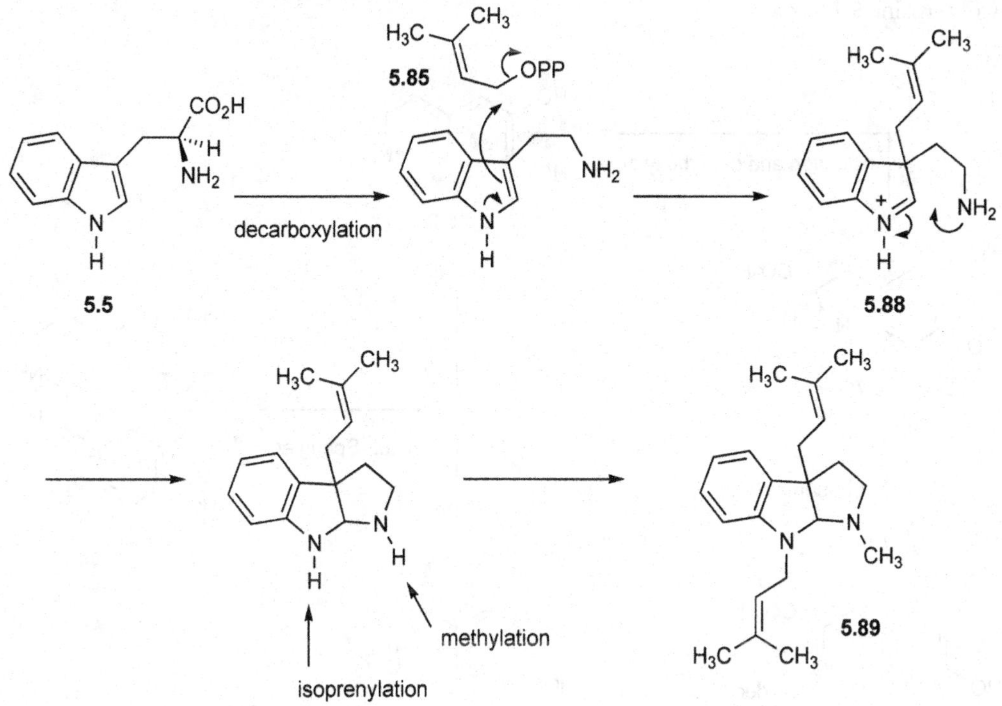

Figure 5.27 Isoprenylation in the formation of alkaloid **5.89**.

Problem

Problem 5.11 The alkaloid eseroline **5.90** is a precursor to the neurologically active compound, physostigmine **5.91**. Propose a reasonable biosynthesis of eseroline **5.90** from tryptophan **5.5**.

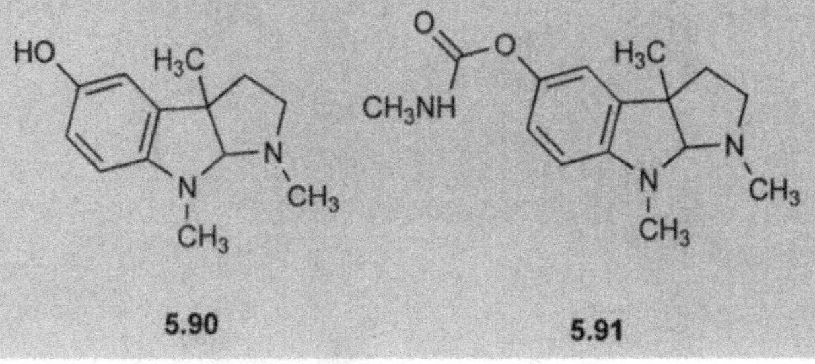

5.5.4 Alkaloids Derived from Tyrosine

Tyrosine **5.7** is the precursor to an immense number of alkaloids. A few representative examples of tyrosine-derived compounds will be considered in this section to illustrate their structural diversity and to show the underlying biosynthetic principles. The formation of norcoclaurine **5.94** is a typical example of how alkaloids possessing a 1,2,3, 4-tetrahydroisoquinoline ring are biosynthesized (Figure 5.28). The amine **5.92** and aldehyde **5.93** fragments which are required for the Pictet–Spengler reaction are derived by oxidation/decarboxylation and transamination/decarboxylation respectively of tyrosine **5.7**.

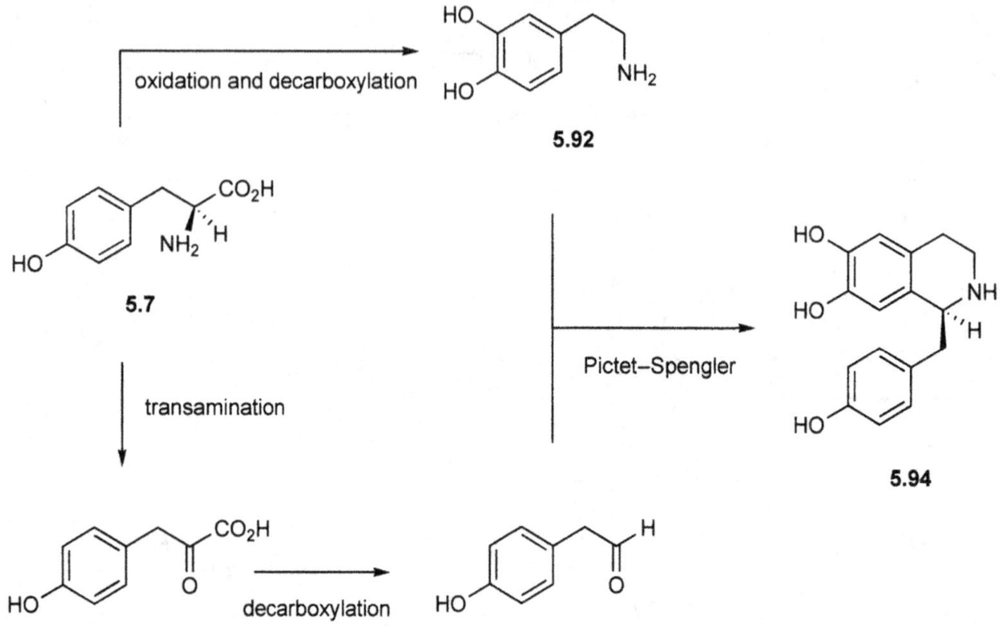

Figure 5.28 Biosynthesis of norcoclaurine **5.94**.

Problem

Problem 5.12 Suggest a reasonable biosynthesis of papaverine **5.95** and (*S*)-reticuline **5.96** from tryrosine **5.7**.

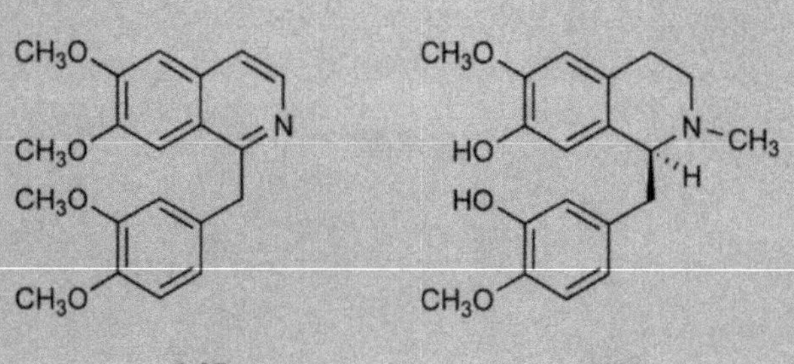

You will notice that alkaloids such as norcoclaurine **5.94** are phenolic. We have seen previously how phenol oxidation gives phenoxy radicals. Similarly, norcoclaurine **5.94** (and other phenol-containing alkaloids) can yield phenoxy radicals which can subsequently dimerize leading to structurally more complex natural products. This concept is illustrated in Figure 5.29 for the formation of thebaine **5.98** from the phenoxy radical **5.97** which is derived by oxidation of (*R*)-reticuline **5.96**. In this example, the unpaired electron at the *para*-position of ring 'A' couples with the unpaired electron at the *ortho*-position of ring 'B' to generate the thebaine skeleton. Thebaine **5.98** is the precursor to morphine **5.99**. Note that in the transformation of thebaine into morphine, one methoxy group is oxidatively removed and one is hydrolytically removed, as depicted in Figure 5.29.

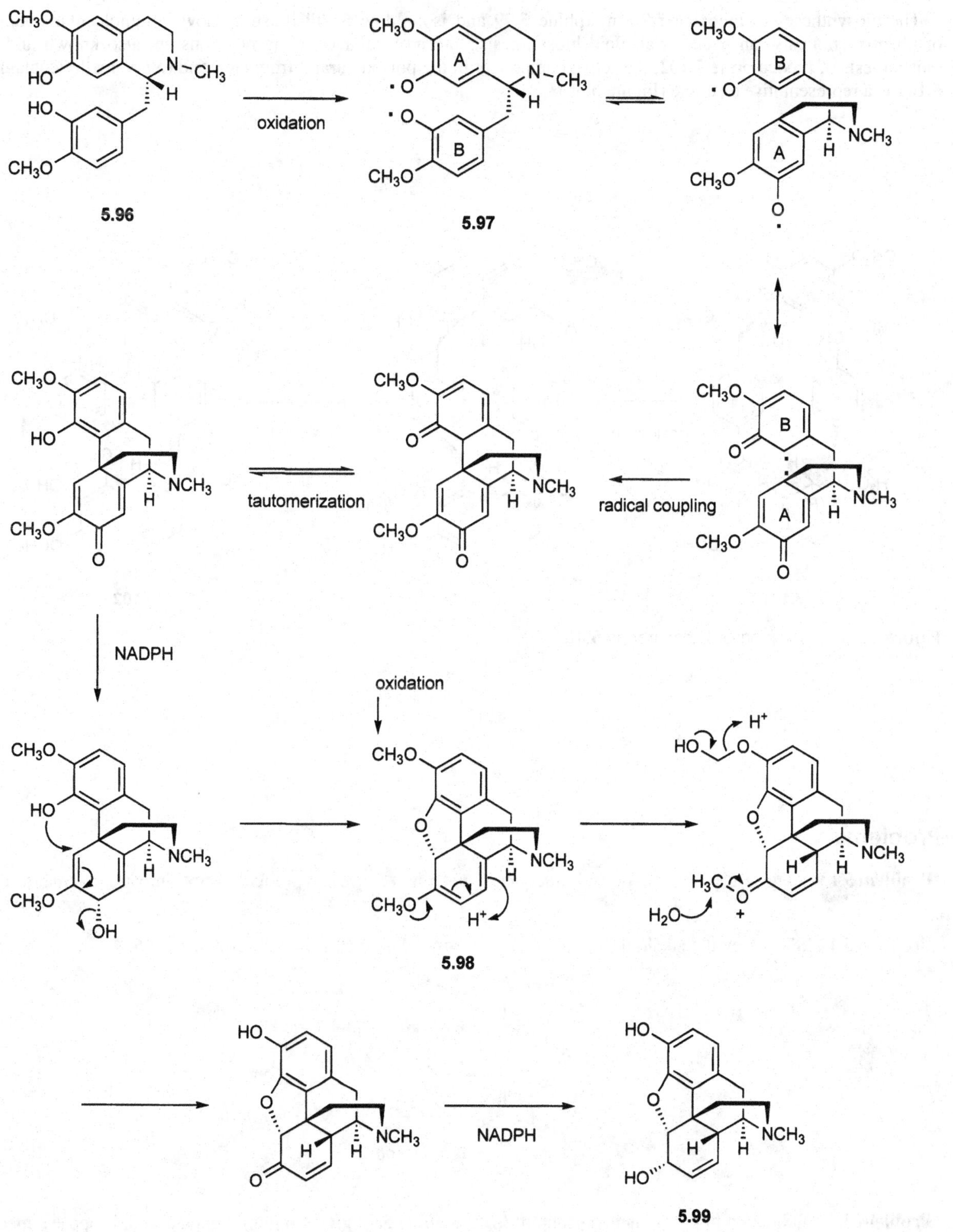

Figure 5.29 Thebaine **5.98** and morphine **5.99**.

The biosyntheses of thebaine **5.98**, morphine **5.99** and isoboldine **5.100** illustrate how intramolecular coupling of phenoxy radicals can occur in alkaloid biosynthesis. Intermolecular coupling reactions are also known and the biosynthesis of tubocurarine **5.102**, the active component in the poison curare, from (*R*)- and (*S*)-*N*-methylcoclaurine **5.101** is a representative example (Figure 5.30).

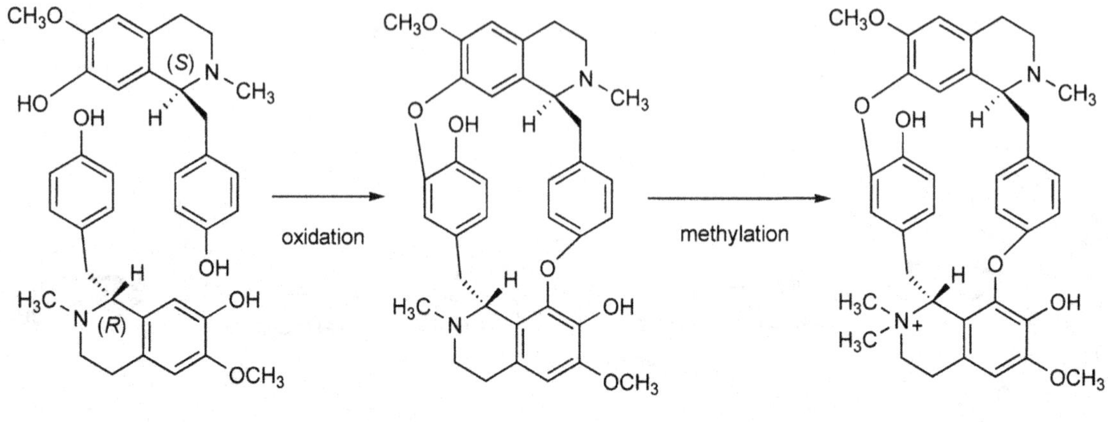

5.101 5.102

Figure 5.30 Formation of tubocurarine **5.102**.

Problems

Problem 5.13 Suggest an oxidation–reduction sequence that might convert (*S*)-reticuline **5.97** into its (*R*)-enantiomer.

Problem 5.14 Show how the alkaloid isoboldine **5.100** might be derived from (*S*)-reticuline **5.96**.

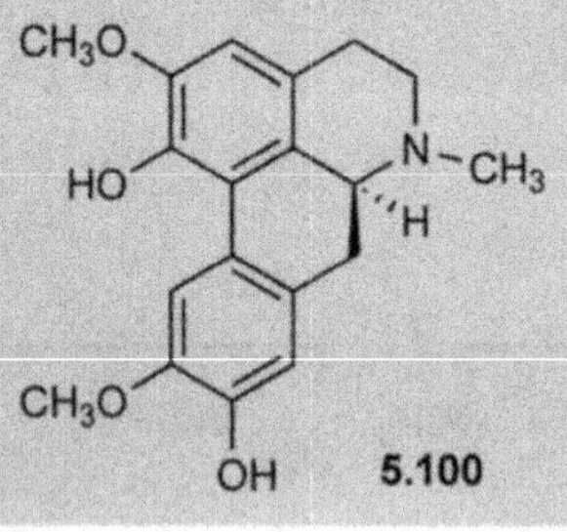

5.100

Problem 5.15 Oxidation of the *N*-methyl group of (*S*)-reticuline **5.97** gives an iminium ion **5.103**. Propose a mechanism for cyclization of this iminium ion to (*S*)-scoulerine **5.104**.

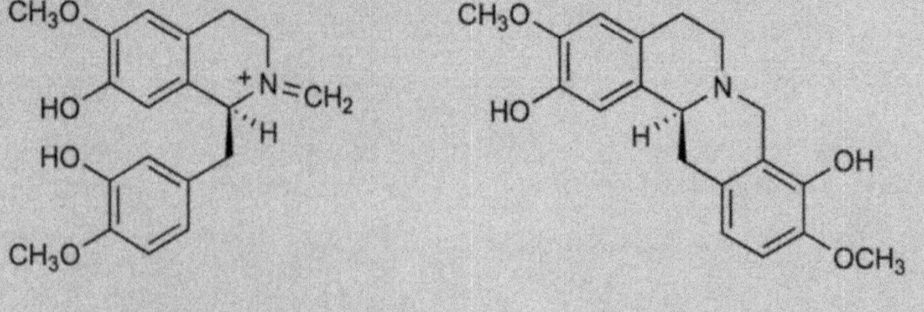

5.103 5.104

Problem 5.16 A proposed biosynthetic route to the alkaloid aaptamine **5.105** is shown below. Suggest a reasonable mechanism for the ring-closure step A (hint: quinones).

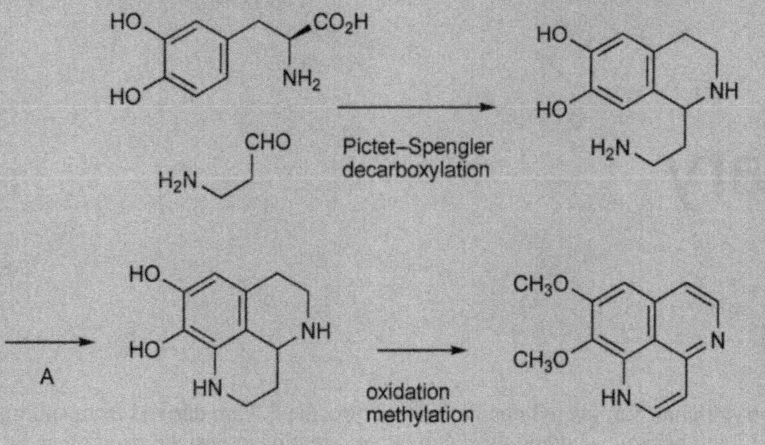

Summary

(a) The amino acids tryptophan **5.5**, phenylalanine **5.6** and tyrosine **5.7** are derived from shikimic acid **5.1**.
(b) Cinnamic acid **5.17** and benzoic acid and their derivatives are biosynthesized from phenylalanine **5.6**.
(c) Cinnamic acid **5.17** and its derivatives are precursors to coumarins, flavones, isoflavones and lignans.
(d) Tryptophan **5.5**, phenylalanine **5.6** and tyrosine **5.7** are precursors to many types of alkaloids.

Terpenes

Objectives

After studying this section you will be able to:

(a) Recognize the role of carbocations in terpene biosynthesis.
(b) Describe how carbocations are formed from pyrophosphate precursors.
(c) Describe important reactions of carbocations in terpene biosynthesis.
(d) Propose reasonable biosynthetic pathways to a variety of terpenes.

6.1 What are Terpenes?

There are a vast number of natural products with a wide range of biological activities that have carbon skeletons possessing 5, 10, 15, 20, 25, ... and so on carbon atoms (Figure 6.1). Examples of C_{10} compounds are the fragrance *trans* citral **6.1** and the insecticide chrysanthemic acid **6.2**; examples of C_{15} compounds are the aphid repellant farnesene **6.3** and the cyclic compound humulene **6.4**; examples of C_{20} compounds are vitamin A **6.5** and the anti-fungal agent sclareol **6.6**. There are relatively few examples of C_5 compounds but the C_5 unit does frequently appear in products of mixed metabolic pathway.

The natural products shown in Figure 6.1 are all members of a class of compounds known as *terpenes* or *terpenoids*. Terpenes are subdivided into groups depending upon the number of carbon atoms as follows:

C_5 hemiterpenes	C_{10} monoterpenes
C_{15} sesquiterpenes	C_{20} diterpenes
C_{25} sesterterpenes	C_{30} triterpenes
C_{40} carotenoids	C_n (n > 40) polyisoprenoids

Some terpenes, most notably *steroids*, do not possess a multiple of five carbon atoms. The steroid cholesterol **6.8** has only 27 carbon atoms and it is biosynthezised from the triterpene lanosterol **6.7** (Figure 6.2). The three methyl groups

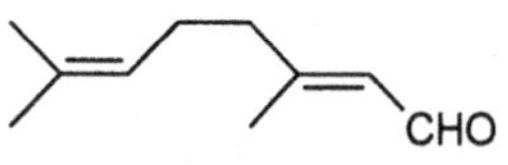

6.1

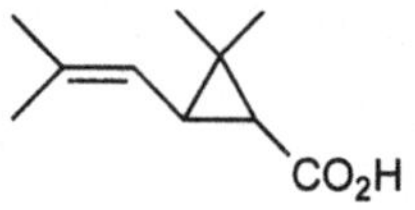

6.2

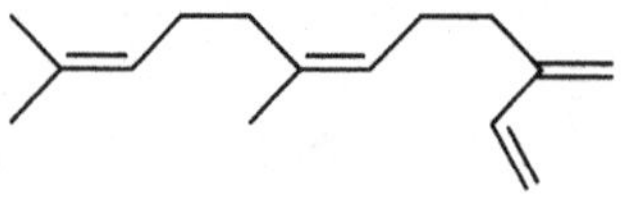

6.3

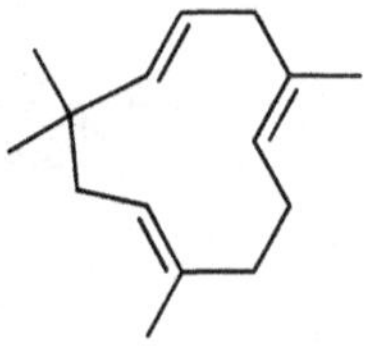

6.4

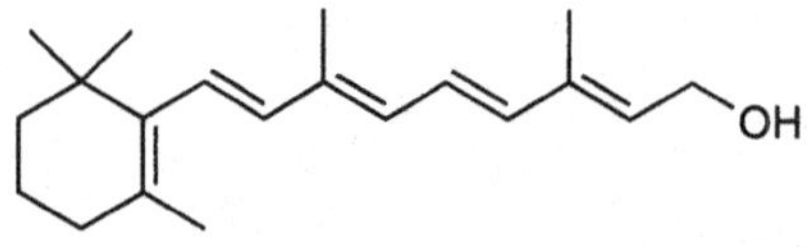

6.5

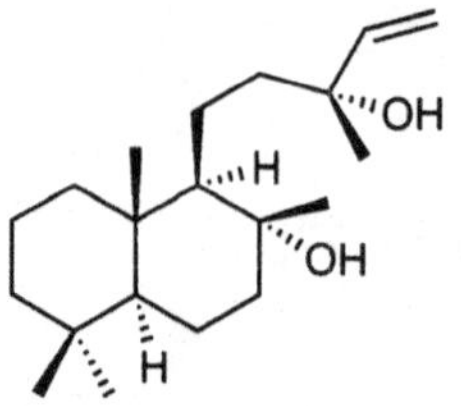

6.6

Figure 6.1 Examples of terpenes.

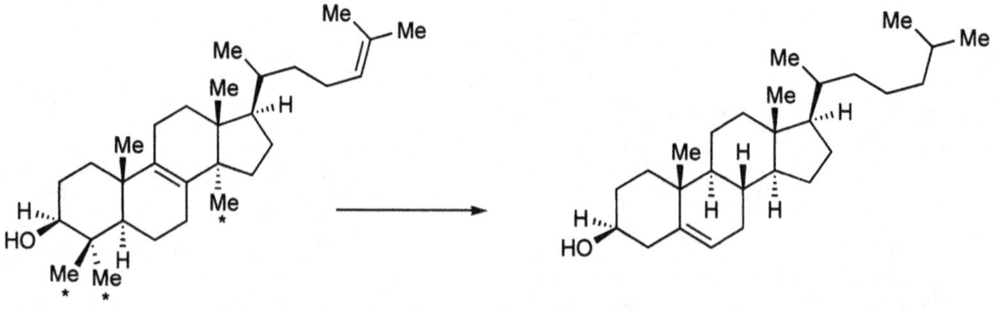

6.7

6.8

Figure 6.2 Lanosterol **6.7** and cholesterol **6.8**.

92

that are starred in Figure 6.2 are lost in the formation of cholesterol **6.8** from lanosterol **6.7**, thus accounting for the 27 carbon atoms of cholesterol.

Problem

Problem 6.1 Terpenes are often refered to as *isoprenoids*, because their carbon skeletons can be drawn as multiples of the C$_5$ unit, isoprene **6.9**. For example, the monoterpene limonene **6.10** contains 2 isoprene units as shown below. Identify the isoprene units in the terpenes shown in Figure 6.1.

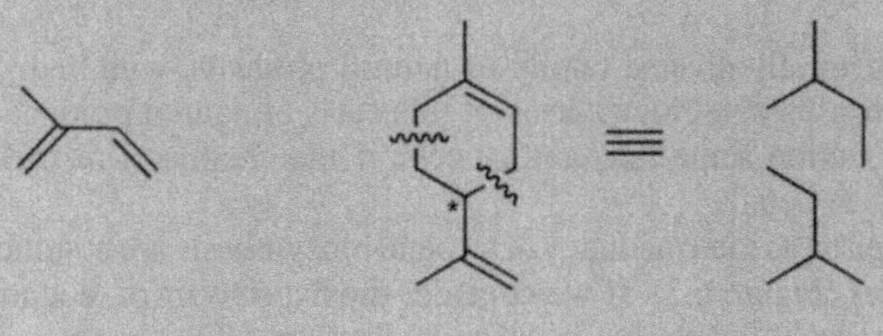

6.9   6.10

Interestingly, many chiral monoterpenes occur in Nature as either enantiomer. For example, limonene **6.10** has one chiral centre (marked with an *) – one enantiomer has the aroma of oranges whereas the other enantiomer smells of lemons.

6.2 Carbocations as Intermediates in Terpene Biosynthesis

Although the terpenes are a structurally diverse family of natural products, with many terpenes also having complex structures, it is reassuring to know that the biosynthesis of this class of natural product can often be rationalized using *carbocation* chemistry. In this section some relevant aspects of the chemistry of carbocations are discussed before considering terpene biosynthesis in detail.

Many of the carbocations required as intermediates in terpene biosynthesis are commonly generated by *heterolysis* of substances called *pyrophosphates* (Figure 6.3). If we consider the heterolysis of alcohols **6.11** giving the carbocations **6.13**, these reactions are not normally favourable because hydroxide (HO⁻) is not a good *leaving group*. In organic chemistry, this problem can usually be circumvented by converting an alcohol function into a good leaving group such as a mesylate group. The mesylate anion ($CH_3SO_3^-$) is far superior to hydroxide as a leaving group. Thus, one method for generating carbocations **6.13** from alcohols **6.11** would be via the mesylates **6.12**. In terpene biosynthesis, the pyrophosphate group is often used to convert alcohols into good leaving groups. Thus, conversion of the alcohols **6.11** into the pyrophosphates **6.14** is used in Nature as a means of generating carbocations **6.13**. The full structure of the pyrophosphate group has been drawn in Figure 6.3, but it is usual to represent the pyrophosphate derivative of an alcohol (R—OH) in the abbreviated form R—OPP. Note that the *curly arrows* in formulae **6.12** and **6.14** indicate heterolysis of the C—O bond; both of the electrons in the σ-orbital of the C—O bond reside on the leaving group resulting in a positively charged carbon atom. The structures of pyrophosphates used in terpene biosynthesis are considered in section 6.4.

Once a carbocation intermediate has been formed by heterolysis of a pyrophosphate it is often converted into an isomeric carbocation leading to considerable structural diversity in terpene biosynthesis. Three particularly important types of reactions of carbocations are: hydride shifts, alkyl shifts, cyclizations.

6.2.1 Hydride Shifts

Figure 6.4 shows a carbocation with the carbon atom labelled *1* bonded to hydrogen. For clarity, no other substituents have been shown on the carbon atoms. If the hydrogen atom migrates from carbon *1* to carbon *2* as shown, this is known as a *1,2-hydride shift*. Since the hydrogen atom migrates with the pair of electrons, it is formally hydride (H⁻) that is migrating. Thus, a new carbocation is formed with the positive charge now centred on carbon atom *1*.

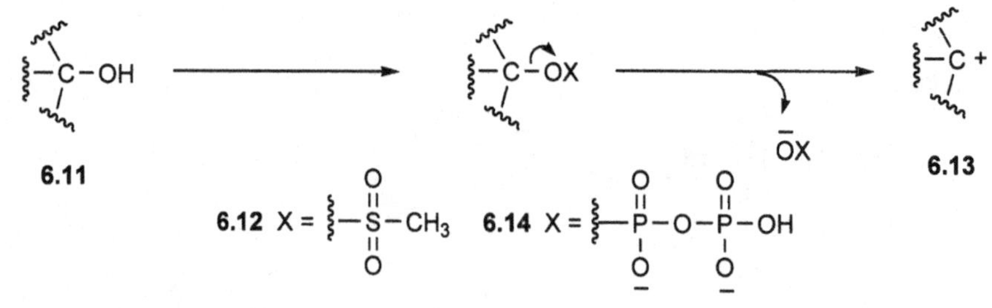

Figure 6.3 Formation of carbocations.

6.2.2 Alkyl Shifts

Figure 6.5 shows a 1,2-methyl shift which is mechanistically similar to the 1,2-hydride shift shown in Figure 6.4. Although a methyl group has been used in this example, other alkyl groups can participate in similar 1,2-alkyl shifts. Such shifts are often described as *Wagner–Meerwein* rearrangements. Quite often these rearrangements can drastically alter the carbon skeleton in terpene biosynthesis. Wagner–Meerwein rearrangements need not involve a discrete carbocation intermediate prior to alkyl-group migration provided that a good leaving group, e.g. a pyrophosphate group is present, as shown in Figure 6.5.

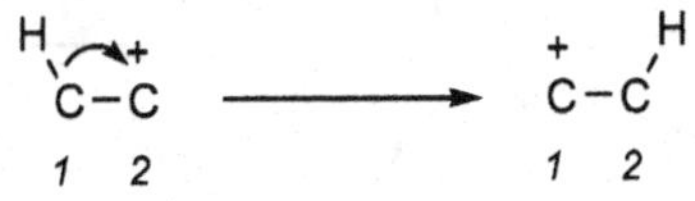

Figure 6.4 Hydride shifts.

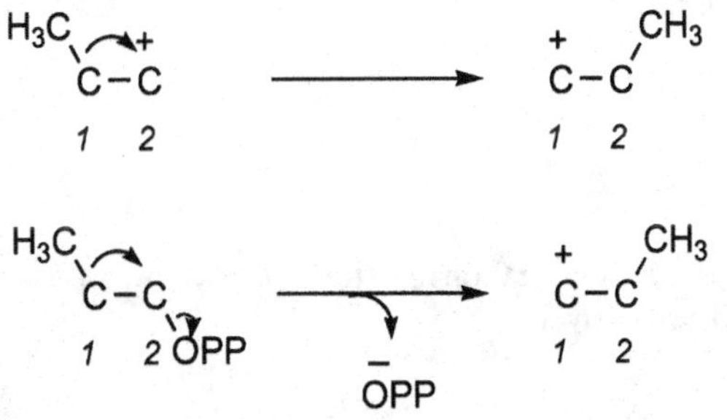

Figure 6.5 Alkyl shifts.

One example of a Wagner–Meerwein rearrangement is shown in Figure 6.6. The carbocation is centred on carbon atom *2* in structure **6.15**, and after rearrangement it is located on carbon atom *1* giving the new carbocation **6.16**. Note that the *1–6* C–C bond has been broken and the *2–6* C–C bond has been formed. Also notice how this alkyl shift has drastically altered the carbon skeleton.

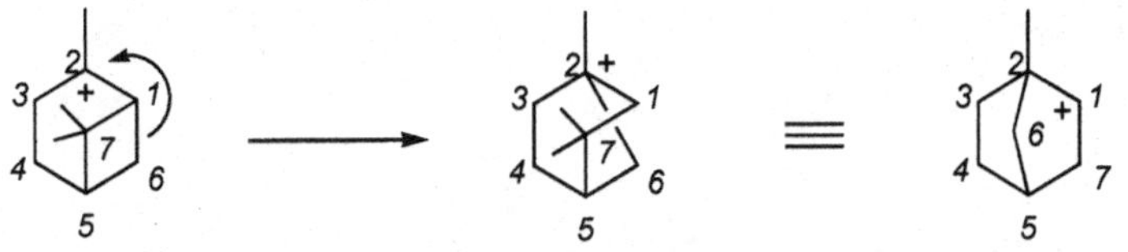

6.15   **6.16**

Figure 6.6 Wagner–Meerwein rearrangement.

6.2.3 Cyclizations

The intramolecular cyclization of carbocationic intermediates and alkene groups is commonly encountered in terpene chemistry. The example shown in Figure 6.7 illustrates how the carbocation **6.17** can be converted into the carbocations **6.15** and **6.18**. Cyclization at the alkene carbon atom labelled *1* gives structure **6.18**, whilst cyclization at the alkene carbon atom labelled *2* gives structure **6.15**.

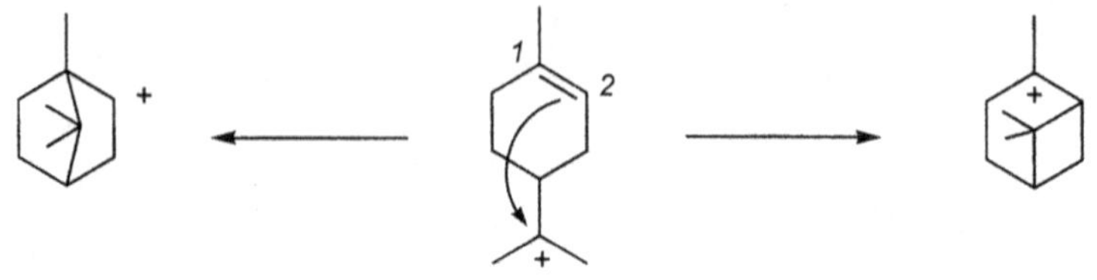

6.18

6.17

6.15

Figure 6.7 Intramolecular cyclizations.

Problem

Problem 6.2 Using 1,2-hydride shifts and cyclization reactions, show how carbocations **6.17** and **6.19** can be converted into the carbocations **6.20** and **6.21** respectively.

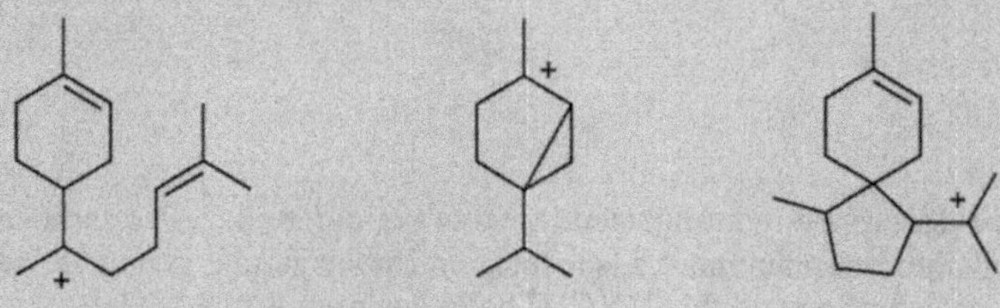

6.19

6.20

6.21

6.3 Termination of Carbocations

Carbocations are intermediates in terpene biosynthesis and there are two major routes through which these carbocations are converted into terpene products. These routes are:

- loss of a proton
- addition of water followed by loss of a proton

6.3.1 Loss of a Proton

Loss of a proton usually occurs from the carbon atom adjacent to the carbocationic centre, giving an alkene derivative. Figure 6.8 shows the biosynthesis of the terpene α-pinene via carbocations **6.15** and **6.17**, in which α-pinene **6.22** is formed from intermediate **6.15** by loss of a proton as shown. You have already met the conversion of carbocation **6.17** into carbocation **6.15** in Figure 6.7. Similarly, limonene **6.10** (see Problem 6.1) is formed from carbocation **6.17** by loss of a proton.

6.3.2 Addition of Water

A carbocation can often undergo nucleophilic addition of water giving, after loss of a proton, an alcohol derivative. The biosynthesis of α-terpineol **6.23** from carbocation **6.17** illustrates this (Figure 6.9).

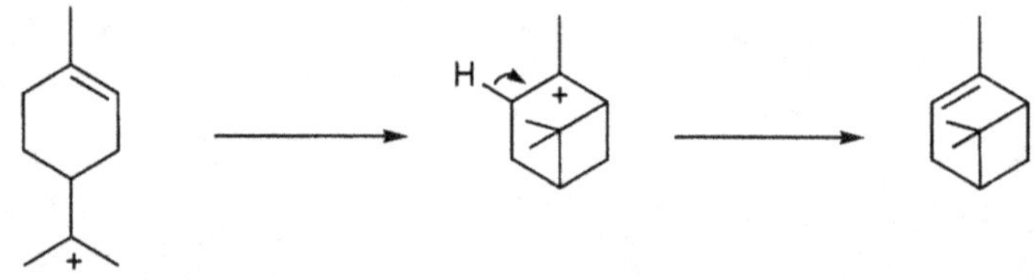

Figure 6.8 Proton loss from carbocations.

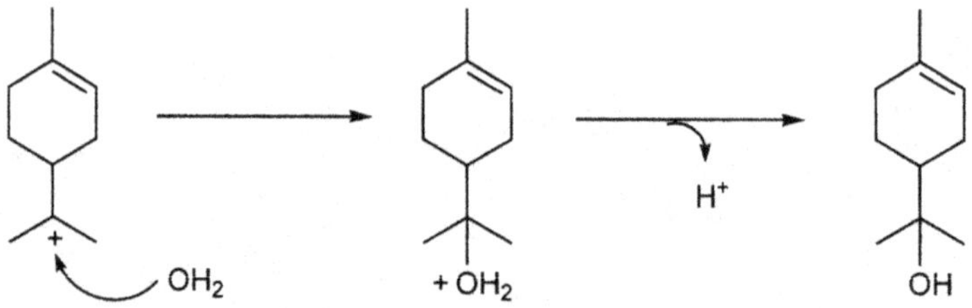

Figure 6.9 Formation of alcohols.

Problem

Problem 6.3 Propose a reasonable biosynthesis of thujene **6.24** from carbocation **6.17** (hint: see Problem 6.2).

6.24

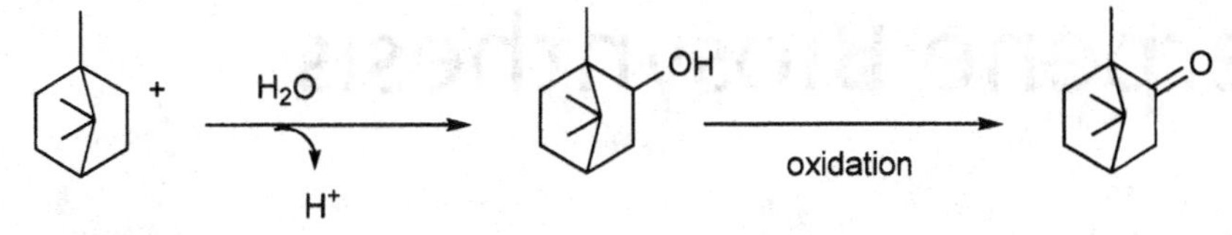

Figure 6.10 Biosynthesis of fenchone **6.25**.

If a secondary or primary alcohol is formed by addition of water to a carbocation, the resulting alcohol can often undergo oxidation. The biosynthesis of fenchone **6.25** from carbocation **6.18** which involves oxidation of a secondary alcohol to a ketone in the final step illustrates this (Figure 6.10).

Many monoterpenes undergo a range of secondary structural modifications, giving a diverse range of natural products. The formation of pulegone **6.26**, menthofuran **6.27**, menthone **6.28** and thymol **6.29** by secondary structural modifications to limonene **6.10** are representative examples (Figure 6.11).

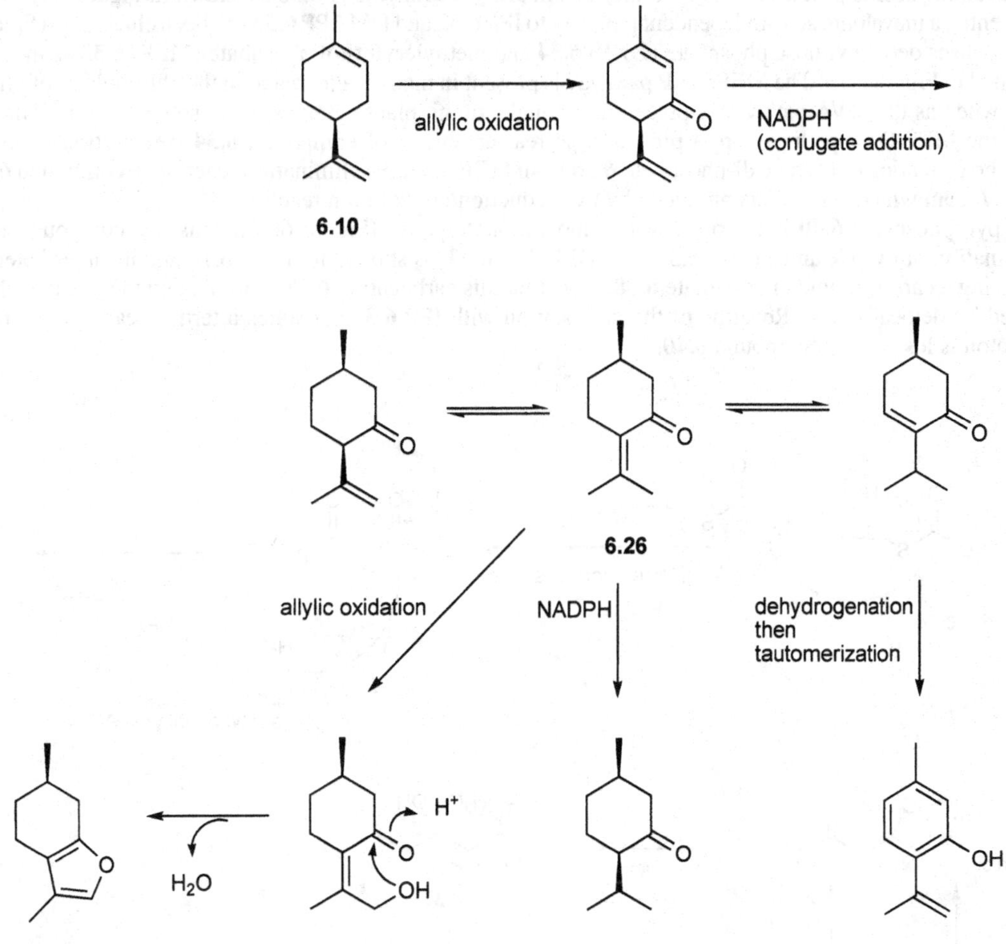

Figure 6.11 Formation of some terpenes from limonene **6.10**.

6.4 Terpene Biosynthesis

6.4.1   Biosynthesis Initiated by Heterolysis of Pyrophosphates

In section 6.2 the formation of carbocations by heterolysis of pyrophosphates was discussed and carbocation **6.17** was used as a typical biosynthetic intermediate to illustrate some common features of carbocation chemistry. The formation of pyrophosphates is now described and hence the sequence pyrophosphate → carbocation → terpene elucidated. The basic building blocks for terpenoids are the two C_5 units, isopentenyl pyrophosphate (IPP) **6.32** and its isomer, dimethylallyl pyrophosphate (DMAPP) **6.33** (Figure 6.12). Two biosynthetic routes to these pyrophosphates have been identified. The first route utilizes mevalonic acid **6.31** as an intermediate and hence is called the mevalonic acid pathway (MVA). Mevalonic acid is produced in several steps from acetyl coenzyme A **6.30** as shown in Figure 6.12.

More recently, a mevalonic acid-independent pathway to IPP **6.32** and DMAPP **6.33** has been elucidated (Figure 6.13). This pathway uses deoxyxylulose phosphate (DXP) **6.34** and methylerythritol phosphate (MEP) **6.35** as intermediates and is outlined in Figure 6.13. The MEP/DXP pathway is present in many bacteria and in the chloroplasts of phototropic organisms, whereas the MVA pathway is present in animals, fungi, plant cytoplasm and some bacteria. The essential features of the MEP/DXP pathway are a pinacol-type rearrangement of compound **6.34** (see section 4, sub-section 4.5.5.) and the formation of a cyclic di-phosphate **6.36** from MEP **6.35**. An elimination reaction in compound **6.36** gives a ketone **6.37** from which IPP **6.32** is produced by two reduction/dehydration reactions.

Geranyl pyrophosphate **6.40** is the precursor to most monoterpenes (Figure 6.14). This C_{10} compound is formed from combination of two C_5 units, (IPP) **6.32** and (DMAPP) **6.33**, as shown in Figure 6.14 and involves heterolysis of DMAPP giving a carbocationic intermediate **6.38**. Note that this carbocation **6.38** is allylic and the positive charge can be stabilized by delocalization. Reaction of the carbocation with IPP **6.32** generates a tertiary carbocation **6.39** from which a proton is lost giving compound **6.40**.

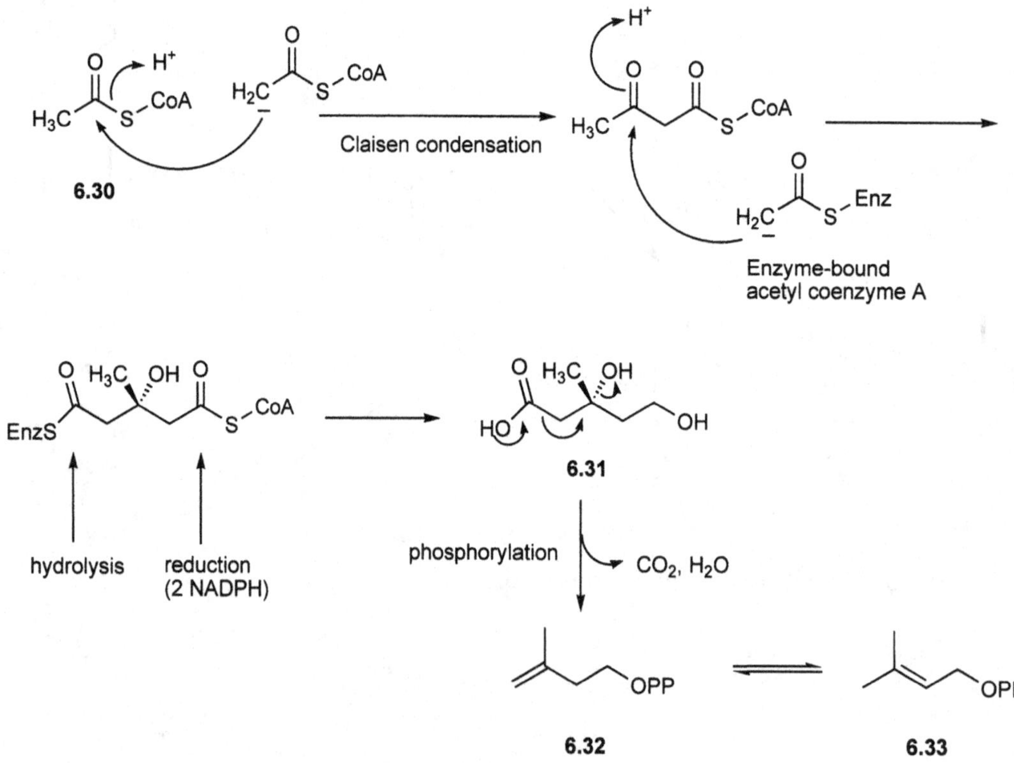

Figure 6.12   Formation of mevalonic acid **6.31**.

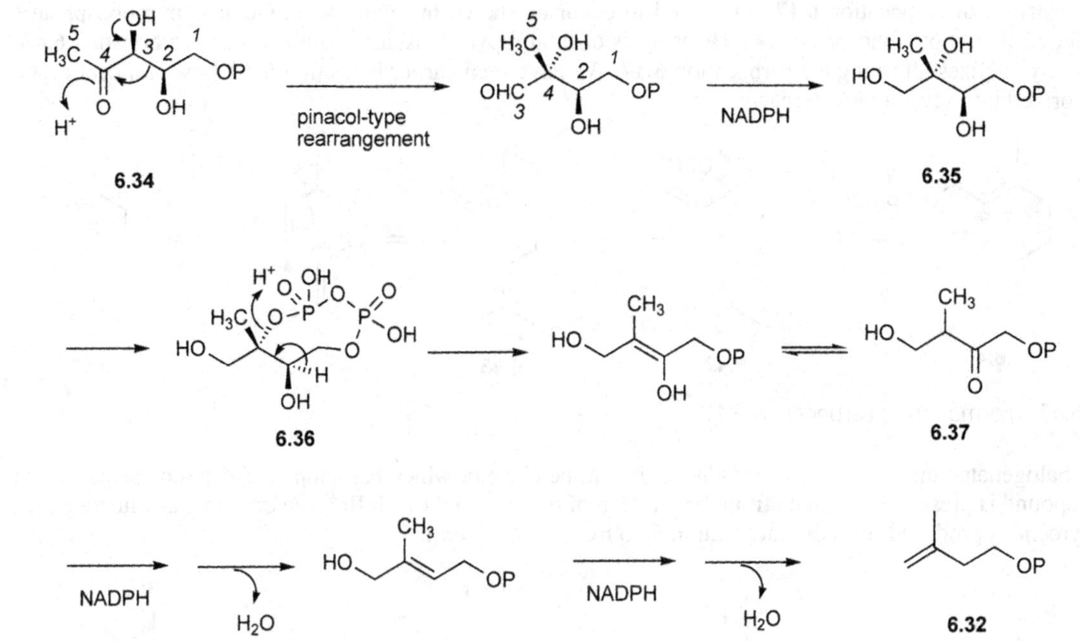

Figure 6.13 Mevalonic acid-independent pathway to terpenes.

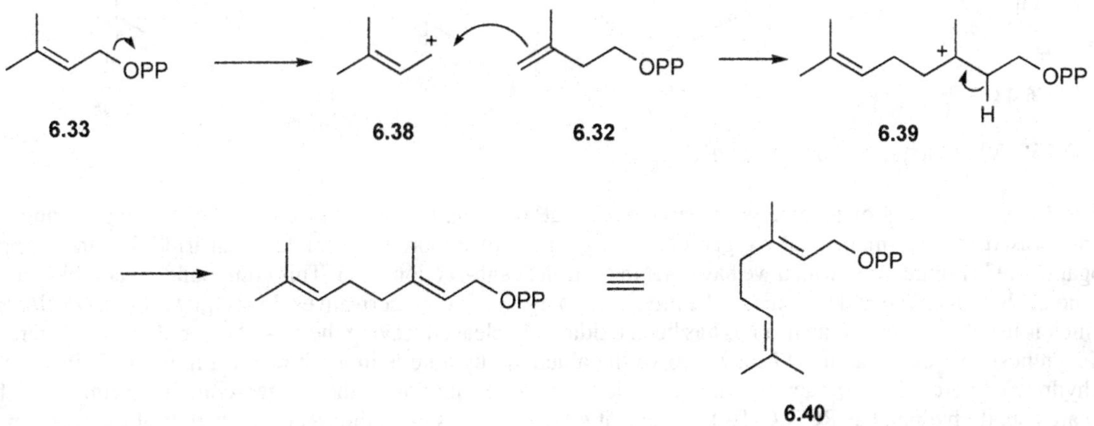

Figure 6.14 Geranyl pyrophosphate **6.40**.

Problem

Problem 6.4 Suggest a reasonable biosynthetic pathway to (i) *trans* citral **6.1** and (ii) the aggregation pheromone **6.41** of the Colorado beetle from geranyl pyrophosphate **6.40**.

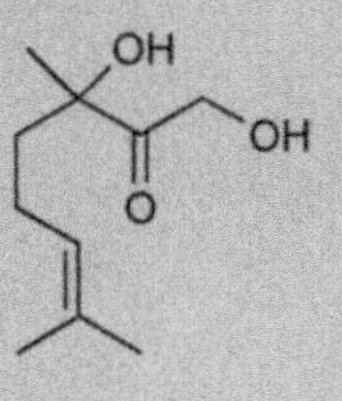

6.41

The formation of carbocation **6.17** is believed to occur as shown in Figure 6.15. Geranyl pyrophosphate **6.40** is isomerized to linalyl pyrophosphate **6.42**. Heterolysis of linalyl pyrophosphate then gives the carbocation **6.43** which subsequently cyclizes affording the carbocation **6.17**. We have seen earlier in section 6.3.2 how carbocation **6.17** can be transformed into several monoterpenes.

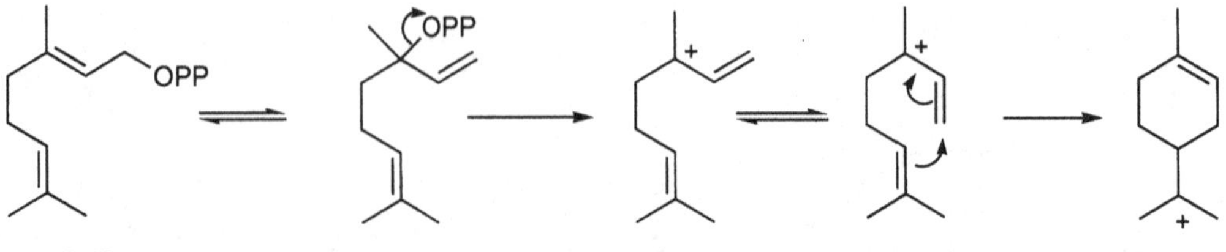

Figure 6.15 Formation of carbocation **6.17**.

Many halogenated monoterpenes are produced by marine algae of which halomon **6.45** is a representative example. This compound is presumably formed from the reaction of myrcene **6.44** with Br^+ and chloride, as illustrated in Figure 6.16. Myrcene is produced from the carbocation **6.43** by loss of a proton.

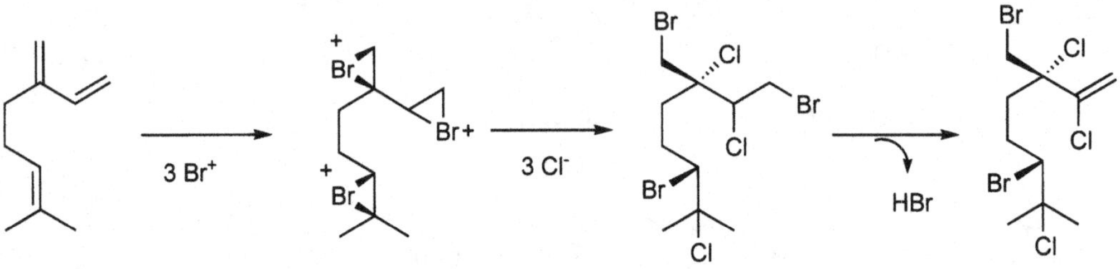

Figure 6.16 A halogenated monoterpene.

The iridoids are a class of monoterpene derivatives that are usually characterized by a cyclopentane ring that is frequently fused to a six-membered oxygen-containing ring. An important example of an iridoid is the compound secologanin **6.47** (Figure 6.17) which we have met in section 5, sub-section 5.5.3. This compound, in combination with the amino acids tryptophan and tyrosine, is the precursor to many alkaloid derivatives. In secologanin, the cyclopentane ring which is usually present in the iridoids has been oxidatively cleaved leaving the six-membered oxygen heterocycle. The biosynthesis of secologanin **6.47** may involve the aldehyde hydrate **6.46** as shown in Figure 6.17, in which one of the hydrate's hydroxyl groups acts as a nucleophile to initiate formation of the oxygen-containing ring. (Aldehydes RCHO are usually hydrated as $RCH(OH)_2$ to an extent which depends upon the electronic nature of the R group.) The oxidative cleavage of the cyclopentane ring may involve fragmentation of an intermediate **6.48**.

Some monoterpenes are *irregular* in their biosynthesis and one example is chrysanthemic acid **6.2**. The formation of the compound is depicted in Figure 6.18 and involves an initial reaction between two molecules of DMAPP **6.33** to form the C_{10} unit, rather than the usual reaction between one molecule each of IPP **6.32** and DMAPP **6.33**. Note how the cyclopropane ring is formed by loss of a proton from a carbon atom which is situated two carbon atoms away from the carbocationic centre. This is a common method for forming cyclopropane rings in terpene chemistry. (Cyclopropane rings can also be formed using this mechanism in fatty acid biosynthesis; for example see section 3, Figure 3.4.) Compare this method of forming cyclopropane rings with the cyclization process you have already encountered with thujene **6.24** in problem 6.3.

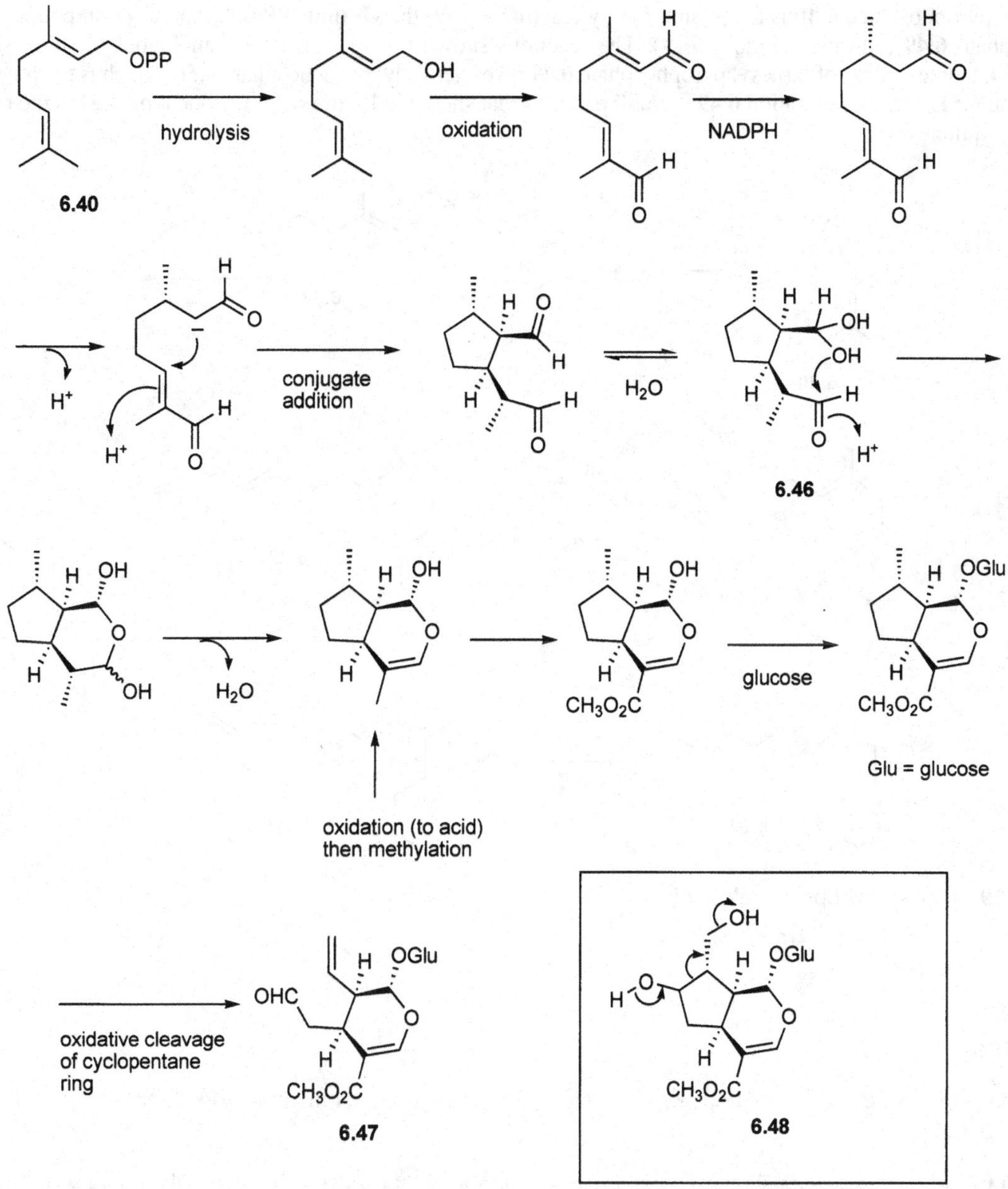

Figure 6.17 Biosynthesis of the iridoid, secologanin **6.47**.

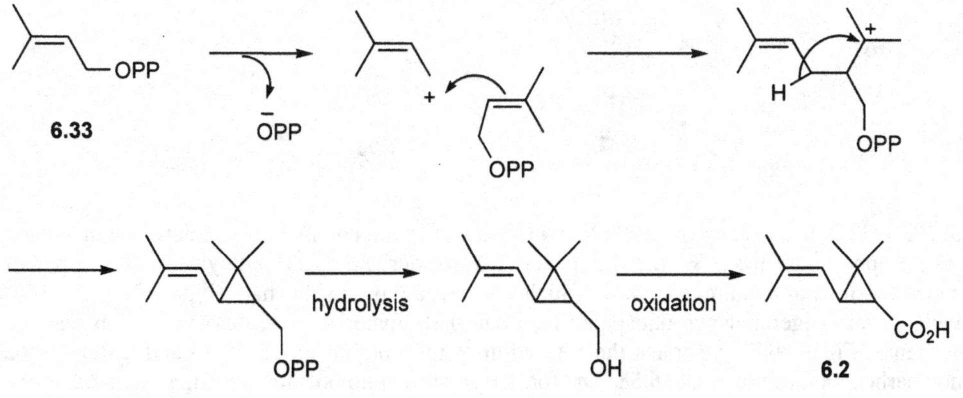

Figure 6.18 Chrysanthemic acid **6.2**: an irregular terpene.

Geranyl pyrophosphate **6.40** is a C_{10} unit and by reacting it with the C_5 unit IPP **6.32**, the C_{15} compound, farnesyl pyrophosphate **6.49** is formed (Figure 6.19). The reactions shown in Figure 6.19 are analogous to those shown in Figure 6.14. Isomerization of farnesyl pyrophosphate **6.49** gives nerolidyl pyrophosphate **6.50**, which is the precursor of carbocation **6.51**. The isomerization **6.49** → **6.50** parallels that shown in Figure 6.14. Carbocation **6.51** is the precursor to many sesquiterpenes.

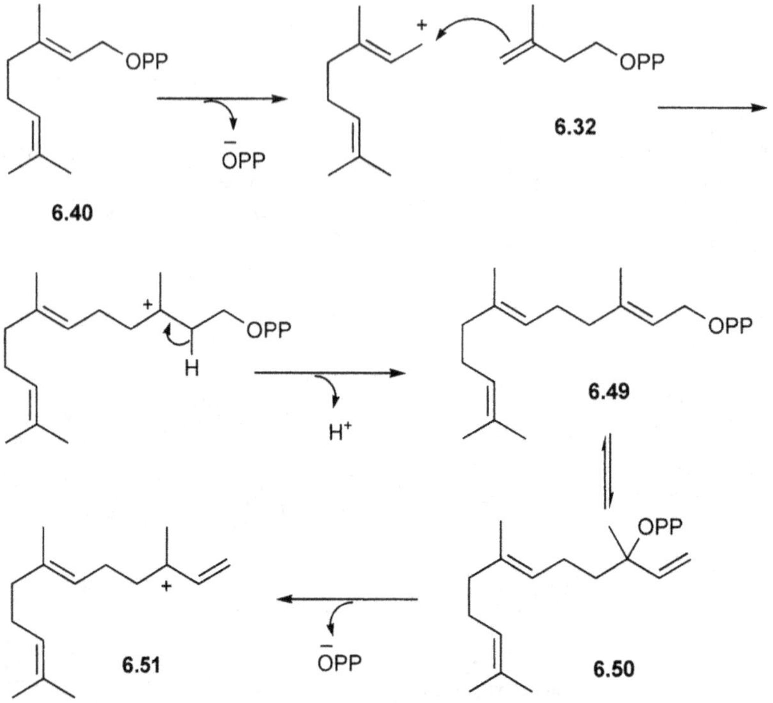

Figure 6.19 Farnesyl pyrophosphate **6.49**.

Problems

Problem 6.5 Show how the sesquiterpenes farnesene **6.3** and humulene **6.4** might be formed from carbocation **6.51**.

Problem 6.6 Show how carbocation **6.19** (see Problem 6.2) might be formed from carbocation **6.51**. How could the sesquiterpenes cedrol **6.52** and α-cedrene **6.53** be biosynthesized from carbocation **6.19**?

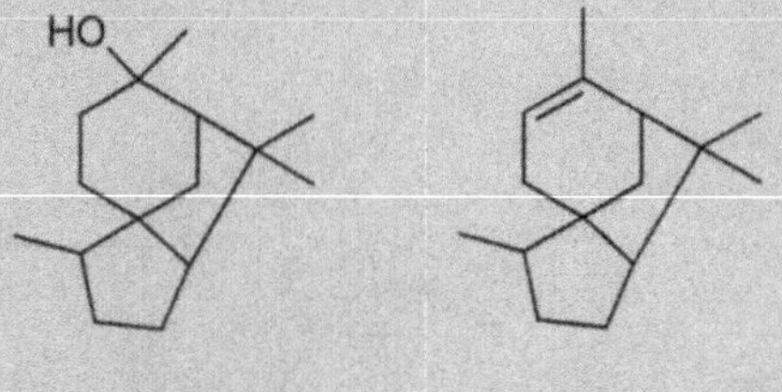

Addition of IPP **6.32** to farnesyl pyrophosphate **6.49** by a similar mechanism to that depicted in Figure 6.19 affords geranylgeranyl pyrophosphate **6.54**. Simple diterpenes that are derived from geranylgeranyl pyrophosphate include phytol **6.55**, plaunotol **6.56** and vitamin K_1 **6.57**, which contains a phytol side chain (Figure 6.20).

Not surprisingly, geranylgeranyl pyrophosphate **6.54** can participate in numerous cyclization reactions yielding a variety of diterpenes. Figure 6.21 illustrates the formation of the phomactatriene **6.59** and taxadiene **6.60** skeletons from a common carbocation intermediate **6.58**. The formation of the phomactatriene ring system **6.59** involves a series of 1,2-hydride and 1,2-methyl migrations. Phomactin B **6.61** is an example of a natural product that possesses the phomactatriene skeleton and this compound is a potent platelet-activating factor. Taxol **6.62**, an anti-cancer compound of contemporary interest, is a well known example of a natural product possessing the taxadiene skeleton.

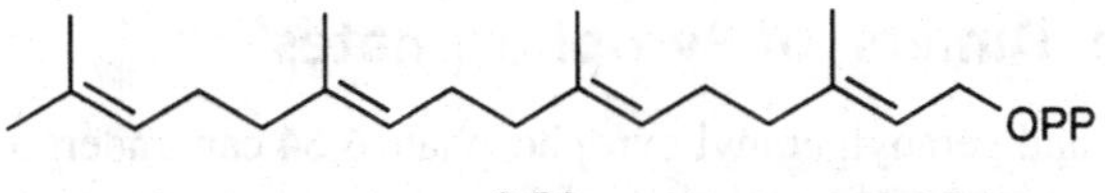

6.54

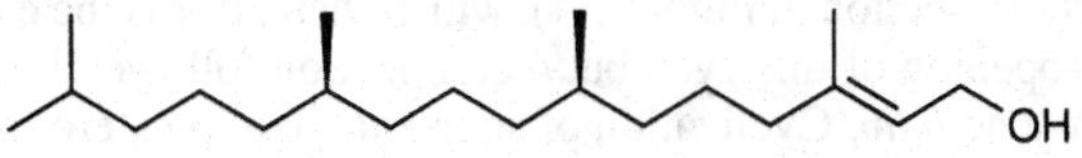

6.55

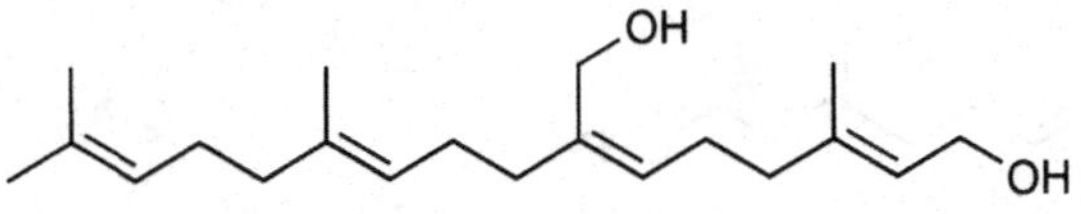

6.56

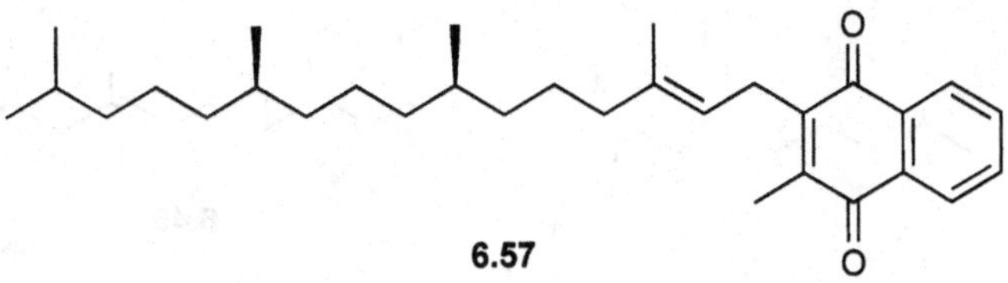

6.57

Figure 6.20 Geranylgeranyl pyrophosphate **6.54**.

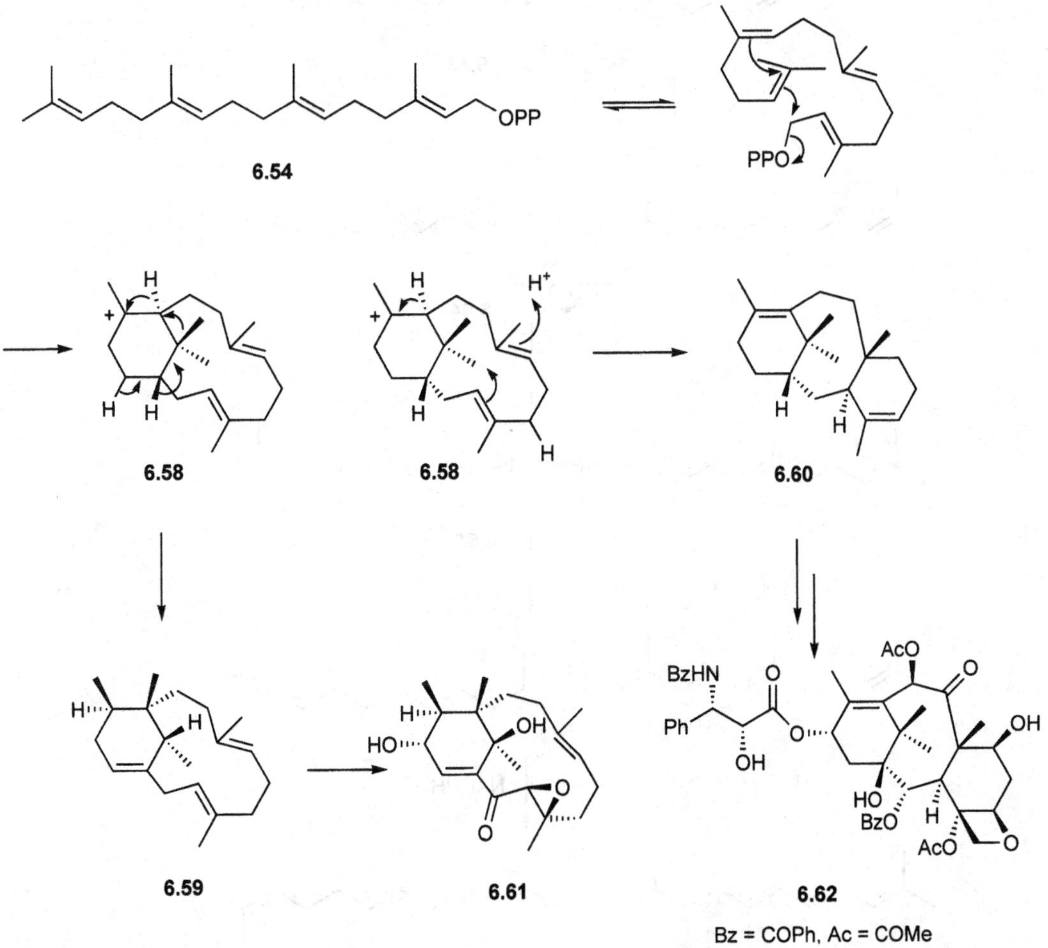

Figure 6.21 Biosynthesis of phomactin B **6.61** and taxol **6.62**.

6.4.2 Biosynthesis from 'Dimers' of Pyrophosphates

Both farnesyl pyrophosphate **6.49** and geranylgeranyl pyrophosphate **6.54** can undergo *head to head* dimerizations, as illustrated in Figure 6.22, for the formation of squalene **6.66** from farnesyl pyrophosphate **6.49**. In the biosynthesis of squalene, loss of a proton from the carbocation **6.63** generates the cyclopropane derivative **6.64** (we have seen related cyclopropane ring formation reactions in section 3, Figure 3.4), which then affords the cyclobutyl carbocation **6.65** by a 1,2-alkyl migration as shown. Ring-opening of this cyclobutyl carbocation followed by reaction of the resulting linear carbocation with NADPH gives squalene **6.66**. Cyclization of squalene yielding steroids and triterpeneoids is initiated either by epoxidation or protonation of the terminal alkene group, and this is considered in section 6.4.3.

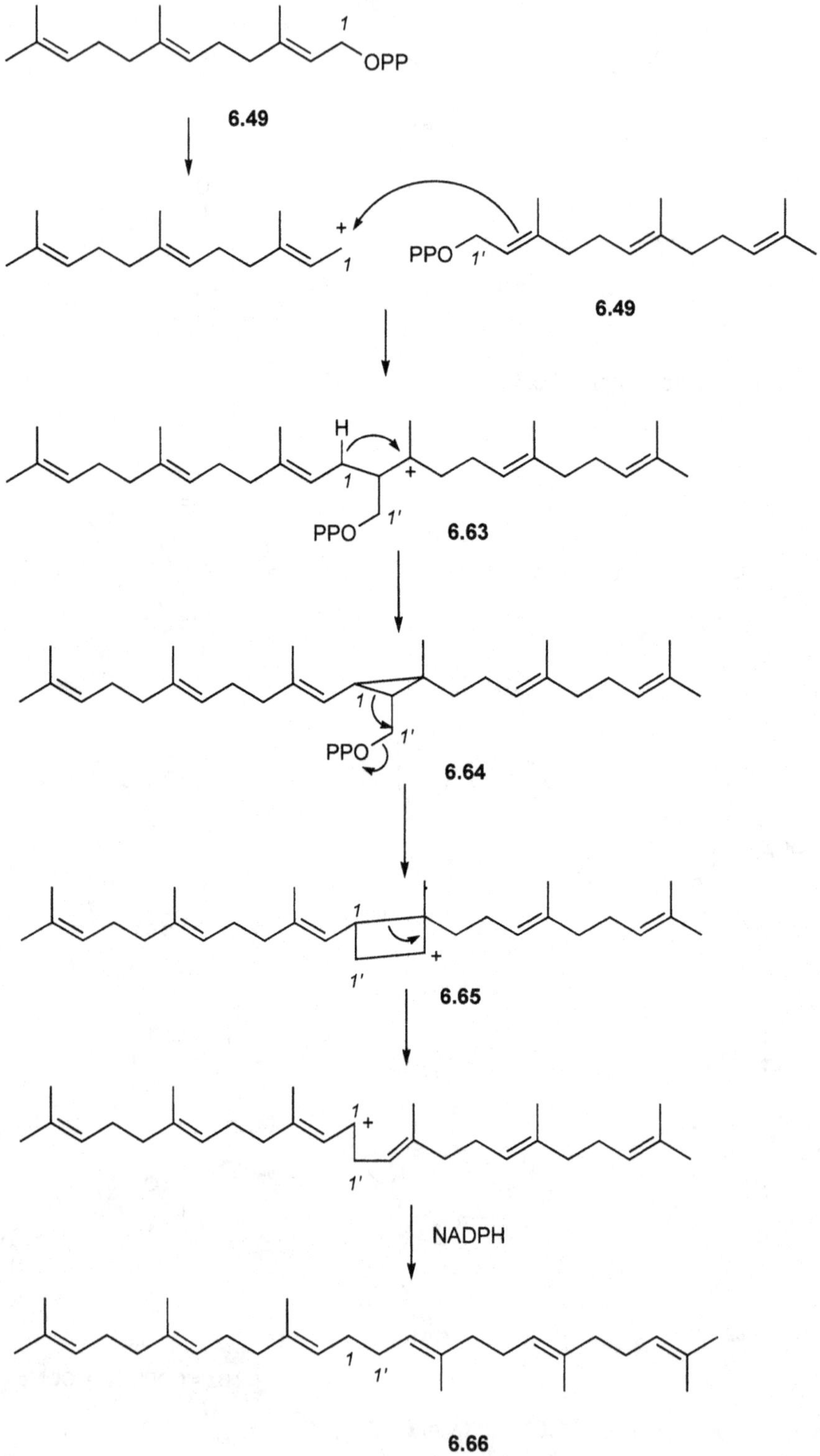

Figure 6.22 Formation of squalene **6.66**.

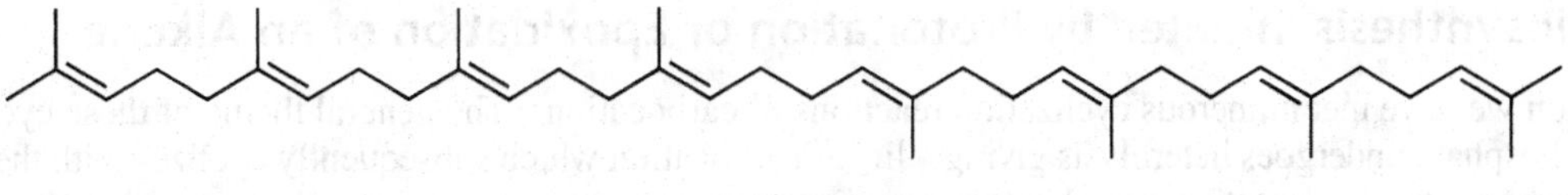

Figure 6.23 Formation of lycopersene **6.67**.

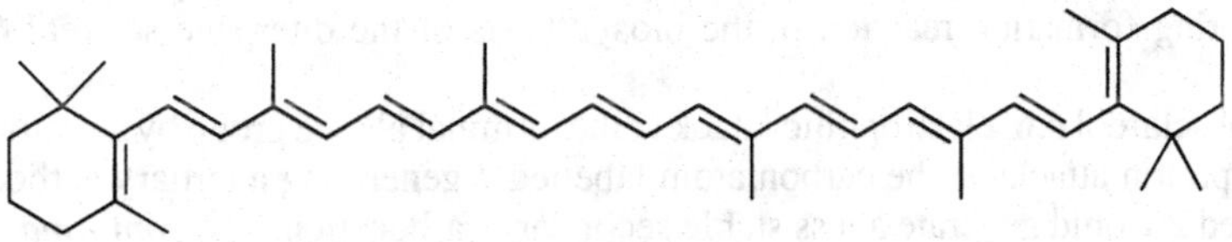

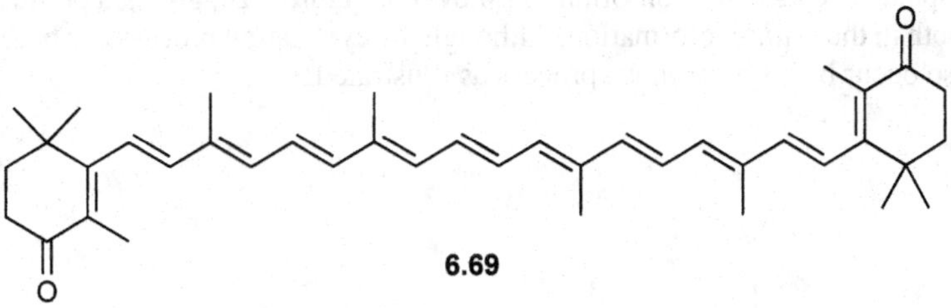

Similarly, geranylgeranyl pyrophosphate **6.54** can give lycopersene **6.67** by *head to head* dimerization (Figure 6.23). Lycopersene is the precursor to highly conjugated pigments such as β-carotene **6.68** and the keto-carotenoids, e.g. canthaxanthin **6.69**. Keto-carotenoids are responsible for the pink coloration in salmon, flamingo feathers and crustacean shells. Oxidative cleavage of β-carotene about the central carbon–carbon double bond gives vitamin A **6.5**.

6.4.3 Biosynthesis Initiated by Protonation or Epoxidation of an Alkene

In this section we have met numerous cyclization reactions of carbocations. The general theme of these cyclizations is that a pyrophosphate undergoes heterolysis giving a linear carbocation which subsequently cyclizes with the formation of a new carbocation. Carbocationic cyclization reactions can also occur in terpene chemistry that do not involve heterolysis of the pyrophosphate as the initial step. These cyclizations are initiated by the addition of an electrophile at one of the alkene groups generating a carbocation which can then undergo ring formation reactions. Figure 6.24 illustrates this type of ring formation reaction in the biosynthesis of the diterpene sclareol **6.6** from geranylgeranyl pyrophosphate **6.54**.

In the biosynthesis of sclareol **6.6**, electrophilic attack at the terminal alkene group by a proton initiates the cyclization process. Note that the proton attacks at the carbon atom labelled *1* generating a tertiary carbocation – proton attack at the carbon atom labelled *2* would generate a less stable secondary carbocation. The *conformation* in which geranylgeranyl pyrophosphate **6.54** cyclizes giving sclareol **6.6** is important in determining the *stereochemistry* of this product. Cyclization of pyrophosphate occurs in the conformation shown in Figure 6.24, giving a product in which the two fused cyclohexane rings are both in the *chair* conformation. Although the cyclization process has been shown using a stepwise mechanism, it could also occur by a synchronous process as illustrated.

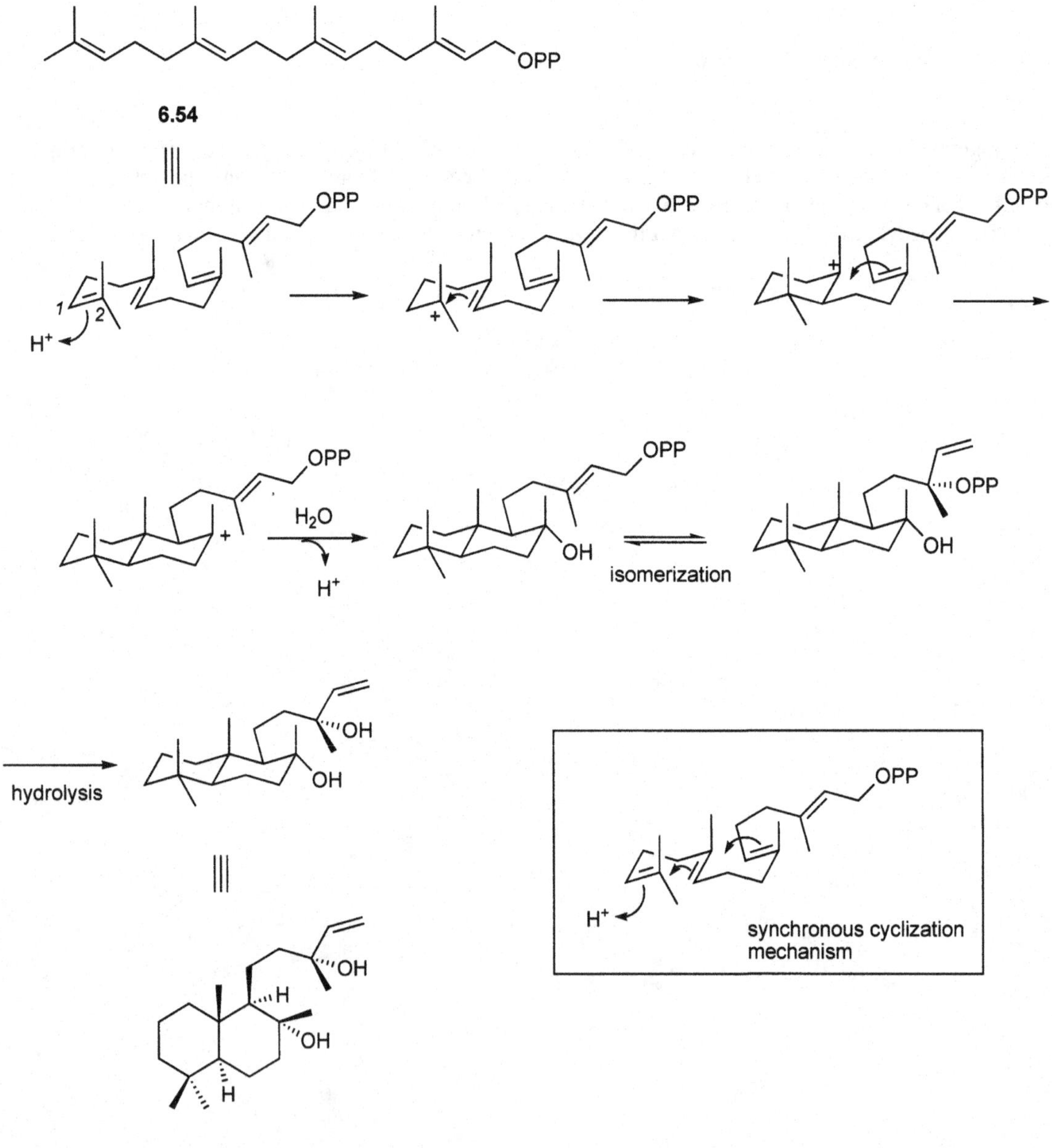

Figure 6.24 Biosynthesis of sclareol **6.6**.

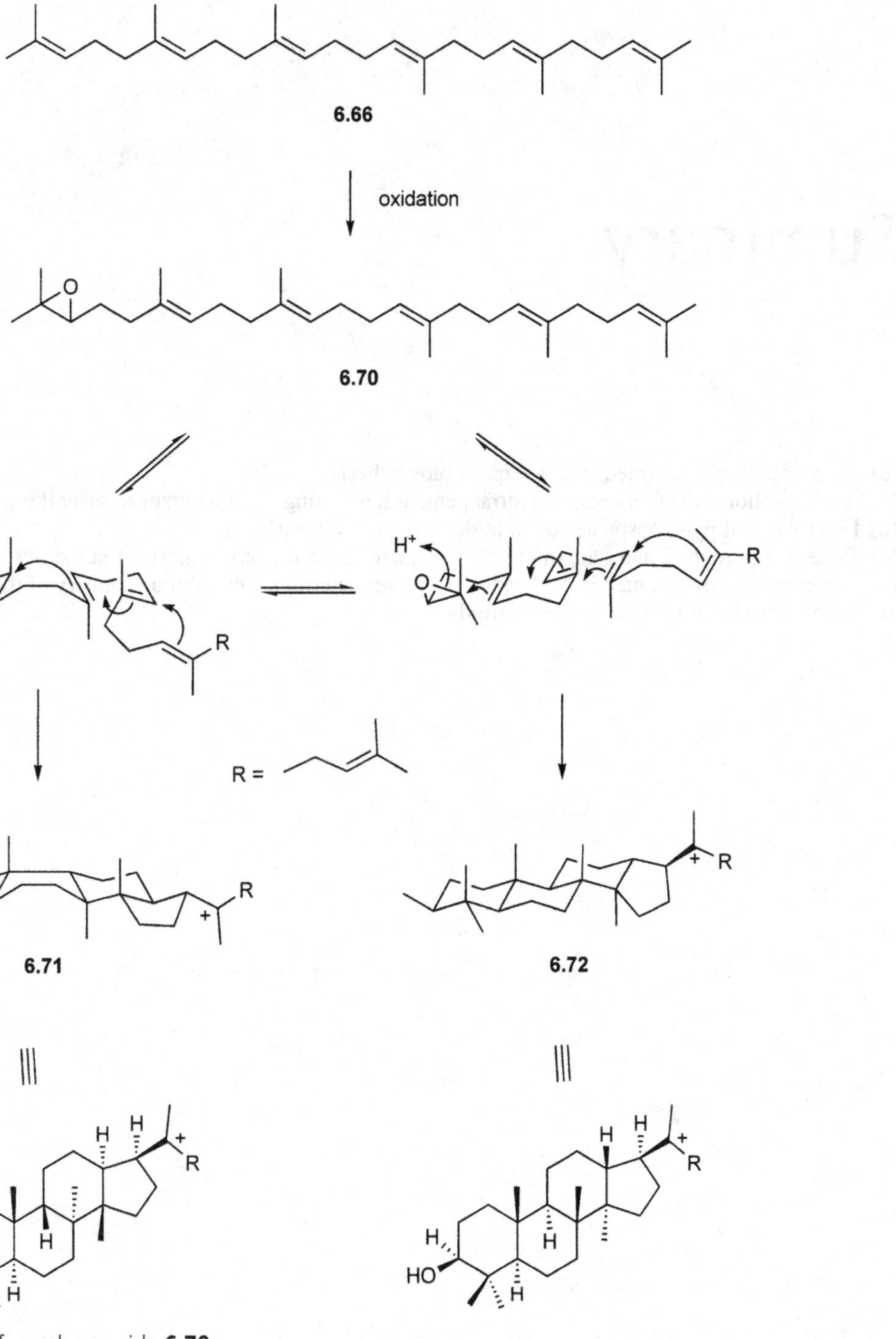

Figure 6.25 Reactions of squalene oxide **6.70**.

One other very important type of cyclization reaction in terpene biosynthesis involves squalene oxide **6.70** (Figure 6.25). Squalene oxide is formed by *epoxidation* at the terminal alkene group of squalene **6.66**. Epoxidation is a well known organic reaction and alkenes can be converted into epoxides by treatment with peracids (RCO_3H) and many other types of oxidizing reagents. The conformation adopted by squalene oxide determines the stereochemistry of the product; both protosterol carbocations I **6.71** and II **6.72** can be formed from squalene oxide **6.70** as shown in Figure 6.25. Note the *chair–boat–chair* conformations of the cyclohexane rings in carbocation **6.71** and the all chair conformations in carbocation **6.72**.

Problem

Problem 6.7 Show how lanosterol **6.7** can be formed from protosterol carbocation I **6.71** using a series of 1,2-hydride and 1,2-alkyl shifts.

Summary

(a) Carbocations are intermediates in terpene biosynthesis.
(b) A carbocation can often undergo rearrangement thus acting as a precursor to several types of terpenes.
(c) Heterolysis of pyrophosphates often initiates terpene biosynthesis.
(d) *Dimers* of pyrophosphates are important intermediates in the biosynthesis of numerous terpenes.
(e) The conformation adopted in cyclization processes determines the stereochemistry of the product.
(f) Squalene oxide is the precursor to steroids.

Natural Products Derived from Amino Acids

Objectives

After studying this section you will be able to:

(a) Identify biosynthetic routes to aliphatic alkaloids from amino acid precursors.
(b) Describe biosynthetic pathways to β-lactams.
(c) Recognize macrocylic peptides.

7.1 Alkaloids

Many aliphatic alkaloids are formed by modification of the amino acids ornithine **7.1** and lysine **7.2**. Figure 7.1 illustrates the conversion of ornithine **7.1** into the cyclic iminium ion **7.5** via the intermediates, putrescine **7.3** and the amino-aldehyde **7.4**. The mechanism of decarboxylation of amino acids giving amines has been discussed in section 5, sub-section 5.6.1. The reaction that converts compound **7.3** into compound **7.4** is a transamination reaction and its mechanism follows that already discussed in section 5, sub-section 5.3, except that the substrate is now an amine rather than an amino acid and the product is an aldehyde rather than an α-keto acid. Similarly, lysine **7.2** can give the iminium ion **7.6**. These two iminium ions **7.5** and **7.6**, and their corresponding N-methylated derivatives **7.7** and **7.8** are the precursors to a wide range of aliphatic alkaloids. Where N-methylation is involved, methylation of the diamine, e.g. compound **7.3** by S-adenosyl methionine occurs. We have seen some examples of how the iminium ions **7.6** and **7.7** can yield alkaloids in combination with the polyketide pathway (section 4, sub-section 4.6).

The pyrrolizidine class of alkaloids, for example laburnine **7.9**, are characterized by two fused five-membered rings with a bridgehead nitrogen atom (Figure 7.2). The biosynthesis of this alkaloid involves two molecules of ornithine

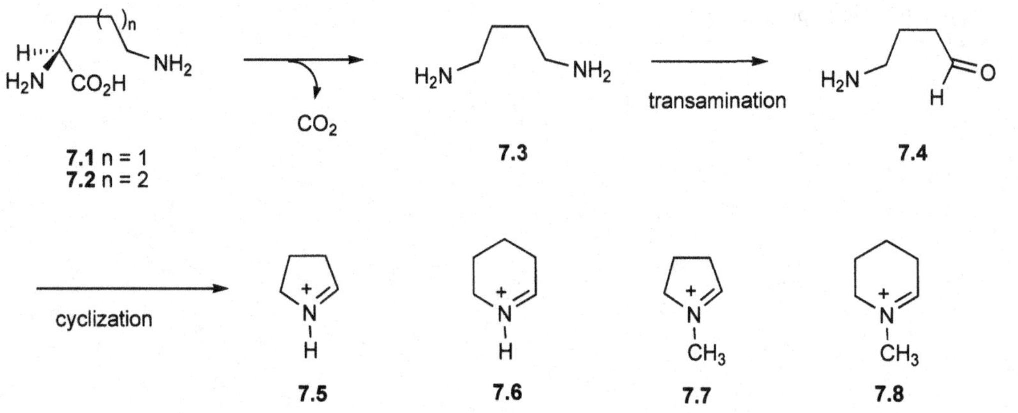

Figure 7.1 Formation of iminium ions.

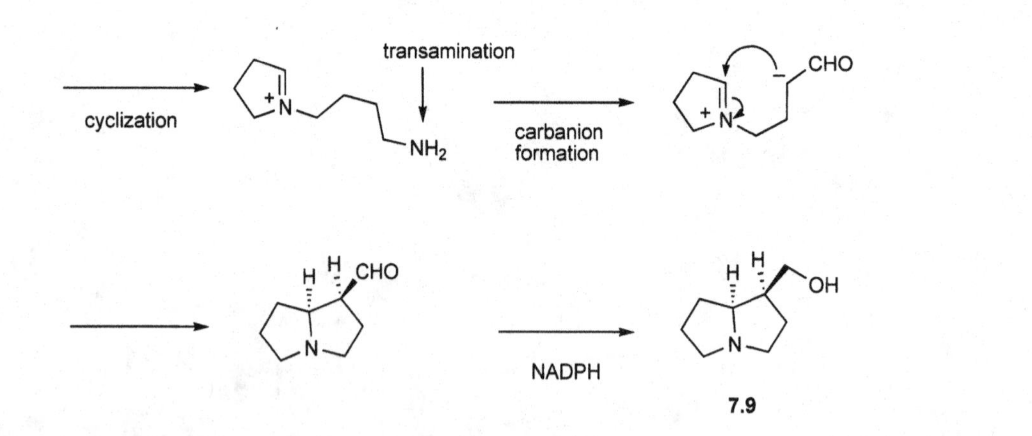

Figure 7.2 Biosynthesis of laburnine **7.9**.

7.1 and uses a series of amine → aldehyde conversions similar to the transformation **7.3** → **7.4** that we have seen in Figure 7.1.

The quinolizidine class of alkaloids possess two fused six-membered rings with a bridgehead nitrogen and are formed from two molecules of lysine **7.2**. Lupinine **7.12** is a representative example whose biosynthesis is outlined in Figure 7.3. In this biosynthesis notice how an iminium ion **7.6** is deprotonated giving an enamine **7.10**: the electron-rich enamine **7.10** can then go on to react with an iminium ion **7.6**. The intermediate aldehyde derivative **7.11** is the precursor of several polycyclic alkaloids of which sparteine **7.13** is an illustrative example (Figure 7.4). The biosynthesis of sparteine **7.13** illustrates nicely the interplay between iminium ion and enamine chemistry.

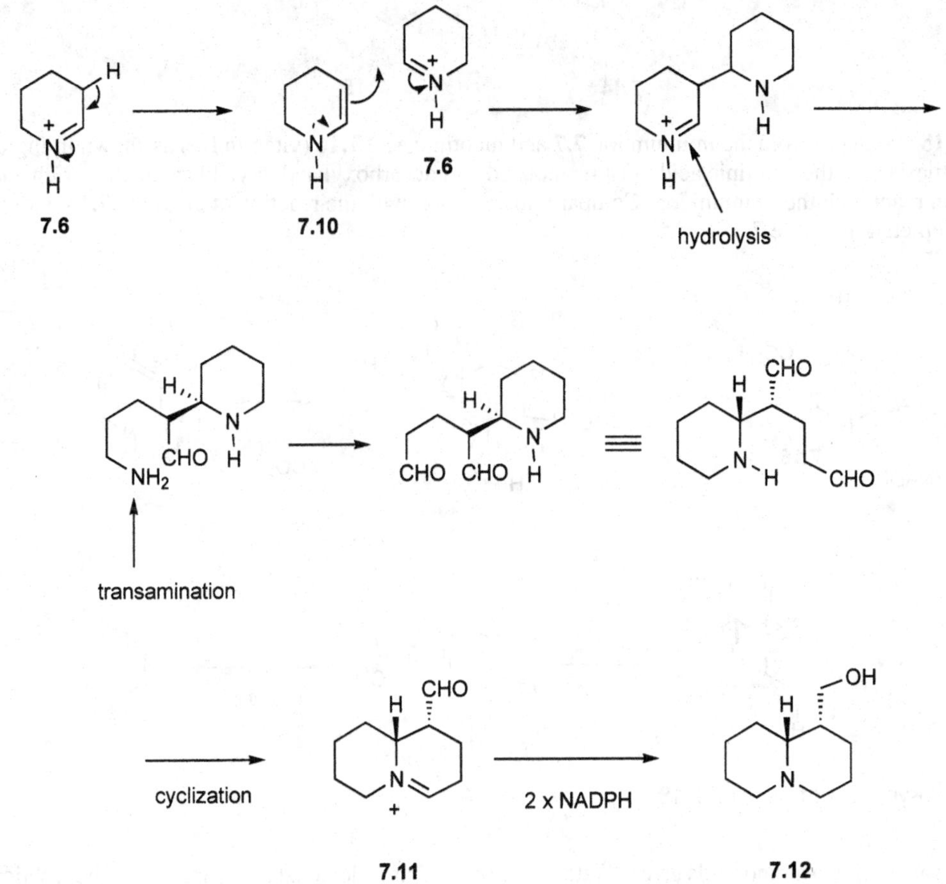

Figure 7.3 Biosynthesis of lupinine **7.12**.

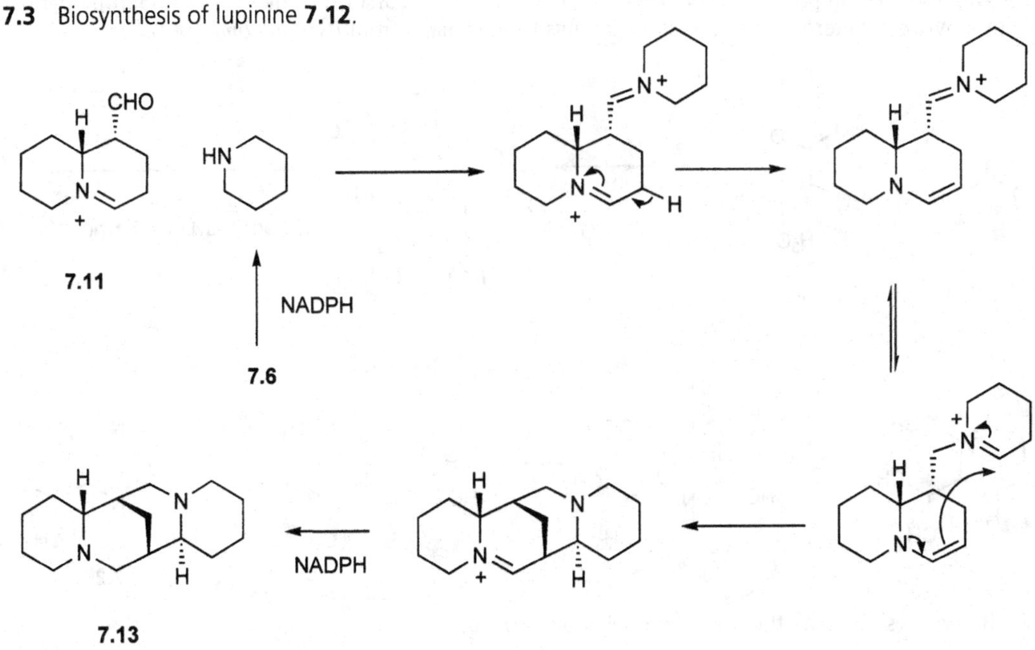

Figure 7.4 Biosynthesis of sparteine **7.13**.

Problem

Problem 7.1 The indolizine class of alkaloids possess a fused five- and six-membered ring with a bridgehead nitrogen atom and compound **7.15** is the key intermediate in their biosynthesis. Suggest how compound **7.15** might be formed from the acetyl coenzyme A derivative of pipecolic acid **7.14** and malonyl coenzyme A. How might pipecolic acid **7.14** be formed from lysine **7.2**?

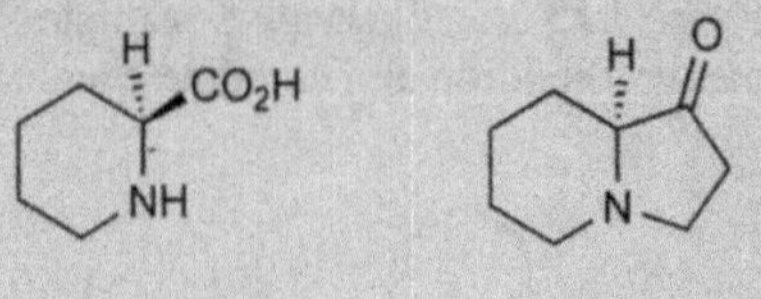

7.14   **7.15**

Nicotine **7.18** originates from the iminium ion **7.7** and nicotinic acid **7.16** (vitamin B3) as shown in Figure 7.5. Note in this biosynthesis how the nicotinic acid must be reduced and decarboxylated to yield an electron-rich enamine **7.17**, which can then react with the iminium ion. Compare this reaction with the reaction of enamine **7.10** with iminium ion **7.6** that was depicted in Figure 7.3.

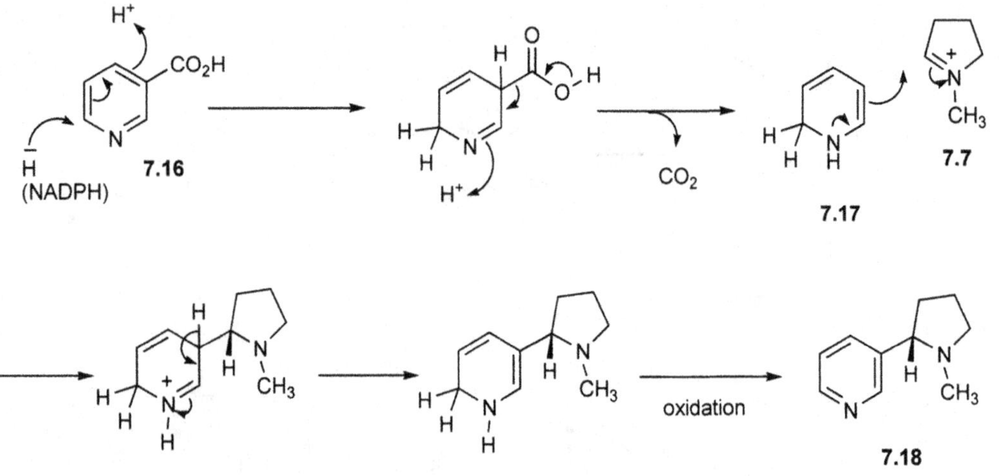

Figure 7.5   Biosynthesis of nicotine **7.18**.

The condensation of two amino acids gives diketopiperazines. This is depicted in Figure 7.6 for the reaction of glycine and valine giving the diketopiperazine **7.19**. This compound is the precursor of 2-methoxy-3-isopropylpyrazine **7.20**, an example of a pyrazine heterocylic ring system that has been isolated from *Pseudomonas perolens*.

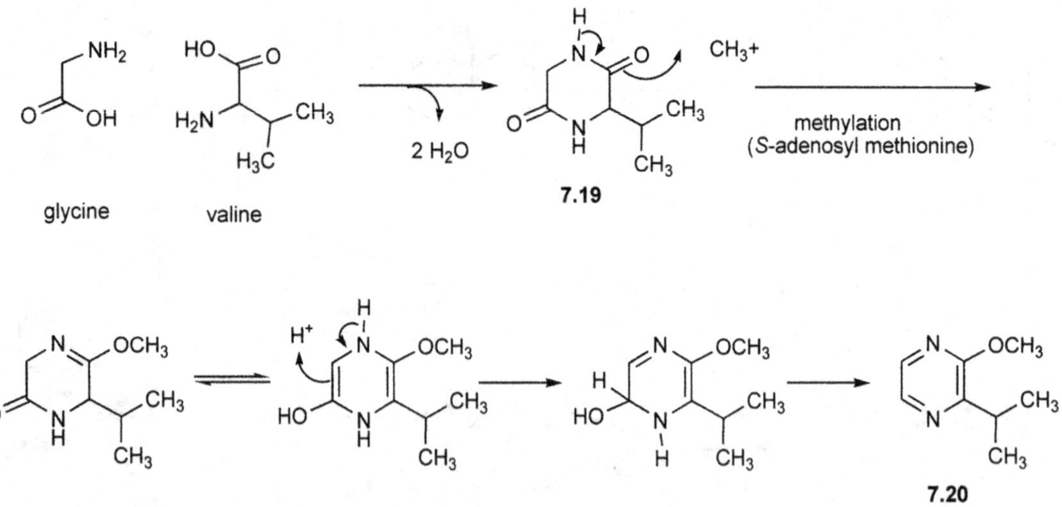

Figure 7.6   Biosynthesis of 2-methoxy-3-isopropylpyrazine **7.20**.

The diketopiperazine skeleton can be seen in many natural products of fungal origin, for example preechinuline **7.21** and mycoediketopiperazine **7.22** (Figure 7.7).

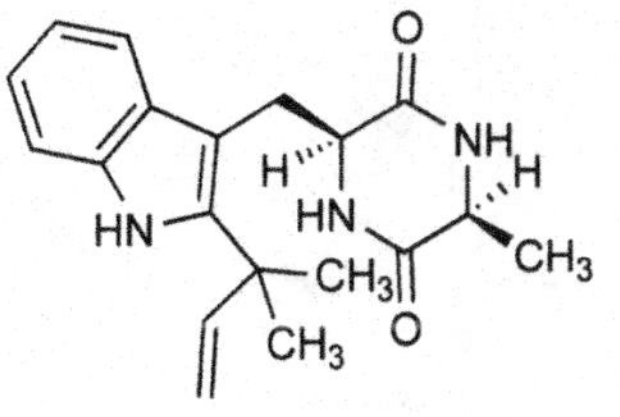

7.21

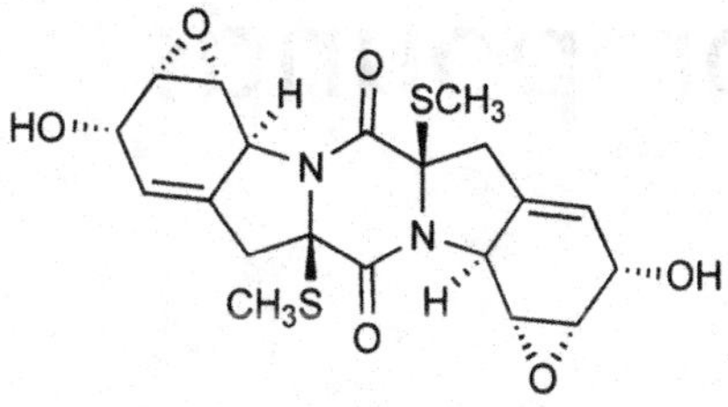

7.22

Figure 7.7   Diketopiperazines.

Problems

Problem 7.2 From which amino acid is compound **7.22** formed?

Problem 7.3 Suggest a plausible biosynthetic route of aspergillic acid **7.23** from two amino acids.

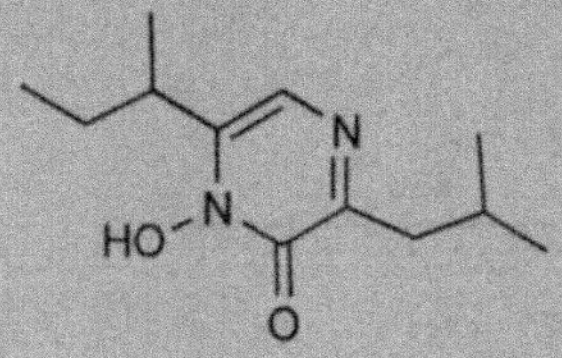

7.23

7.2 Penicillins and Related Compounds

The condensation of three amino acids, aminoadipic acid, cysteine and valine gives the tripeptide **7.24** that is the precursor of the penicillin class of antibiotics (Figure 7.8). The remarkable formation of the bicyclic ring system **7.25** is catalyzed by the enzyme isopenicillin N synthase (IPNS), an oxidase enzyme that requires iron as a cofactor. The mechanism outlined in Figure 7.8 has been proposed for the conversion of the tripeptide **7.24** into compound **7.25**. Note how the oxidation state of the iron changes at various stages in the reaction and how the cyclization is initiated by a peroxy radical, Fe—O—O⋅. Hydrolytic removal of aminoadipic acid yields 6-aminopenicillanic acid **7.26** from which penicillins, e.g. penicillin G **7.27** are obtained.

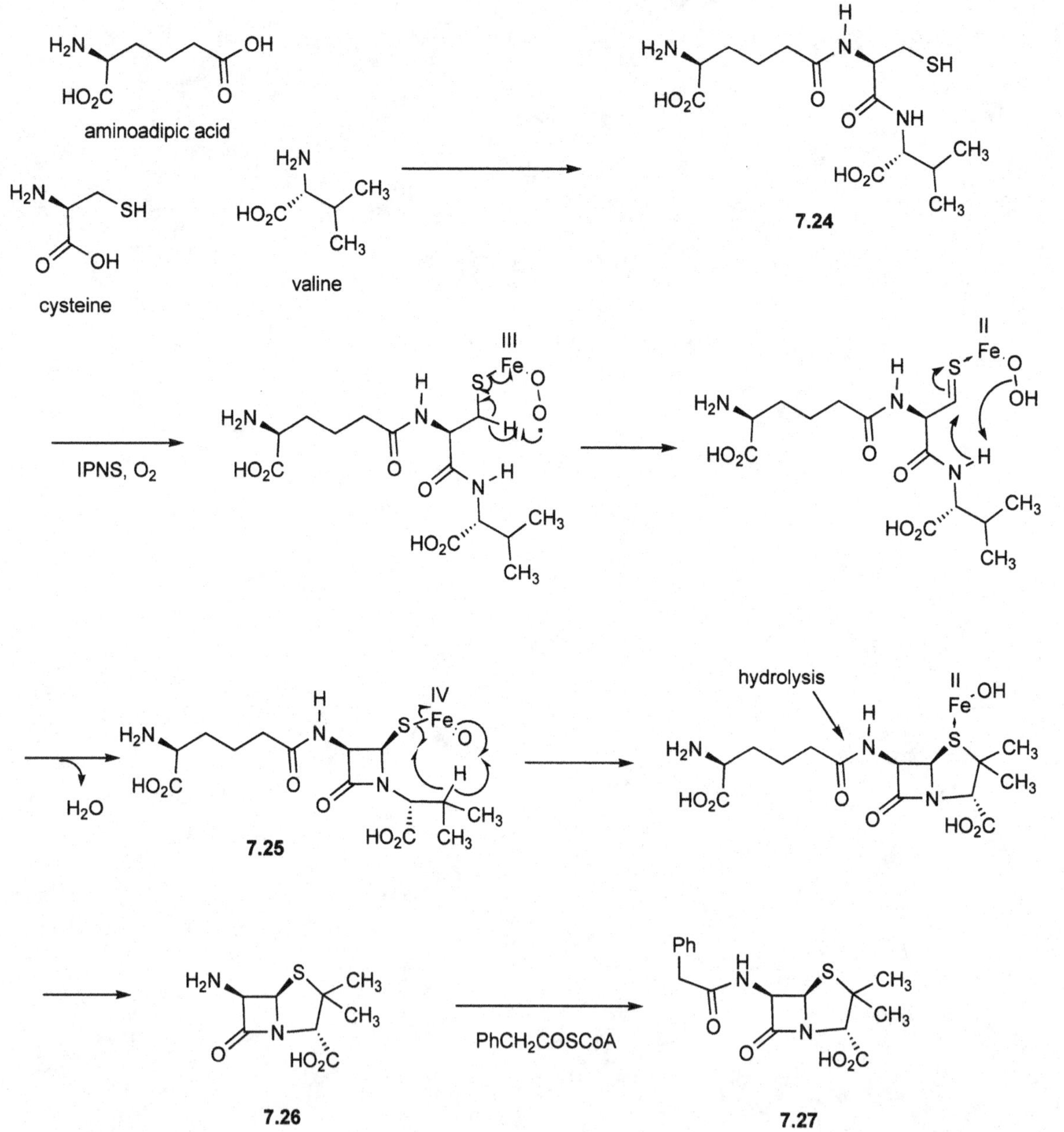

Figure 7.8 Biosynthesis of penicillin G **7.27**.

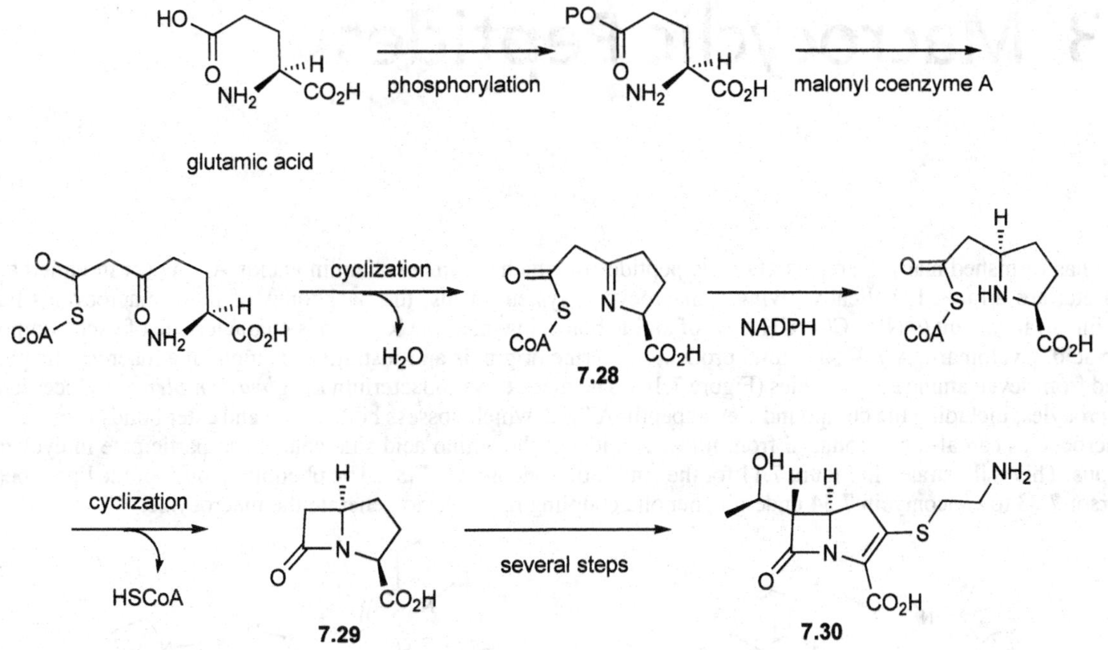

Figure 7.9 Biosynthesis of thienamycin **7.30**.

There are other methods for the formation of the bicyclic ring in compounds that are structurally related to penicillins. For example, thienamycin **7.30** is formed from glutamic acid as shown in Figure 7.9 via the monocyclic **7.28** and bicyclic **7.29** rings. Activation of the carboxylic acid group in glutamic acid is achieved by phosphorylation.

7.3 Macrocyclic Peptides

Nature has furnished many examples of cyclic peptides in which a number of amino acids A_1, A_2 (or modified amino acids) etc. are arranged cyclically giving a macrocyclic system. Thus, the backbones of these macrocycles have a repeating arrangement $(-NH-CO-CHR-)_n$ of amide bonds in which the R group is characteristic of each component amino acid. Cyclomarin A **7.31**, a natural product of marine origin, is an illustrative example of a macrocyclic peptide formed from seven amino acid residues (Figure 7.10). The marine cyanobacterium *Lyngbya semiplena* produces a range of macrocycles, including the compound wewakpeptin A **7.32**, which possess both amide and ester bonds (Figure 7.10).

Macrocycles can also be produced from linear peptides if the amino acid side chains can participate in cyclization reactions. This is illustrated in Figure 7.11 for the antibiotic vancomycin **7.34**. The phenolic groups of the linear peptide precursor **7.33** to vancomycin **7.34** undergo phenolic coupling reactions to generate the macrocycle.

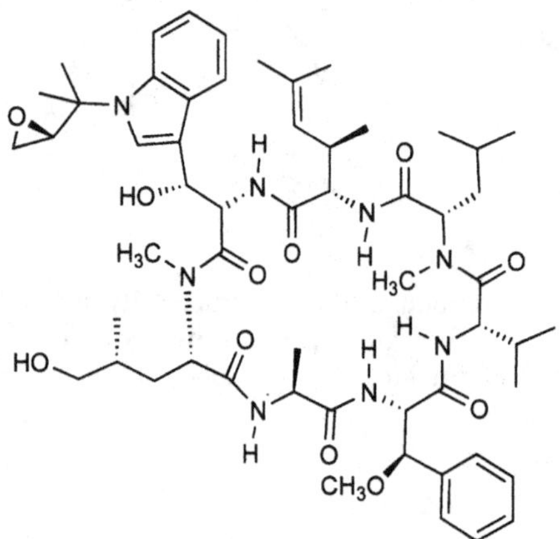

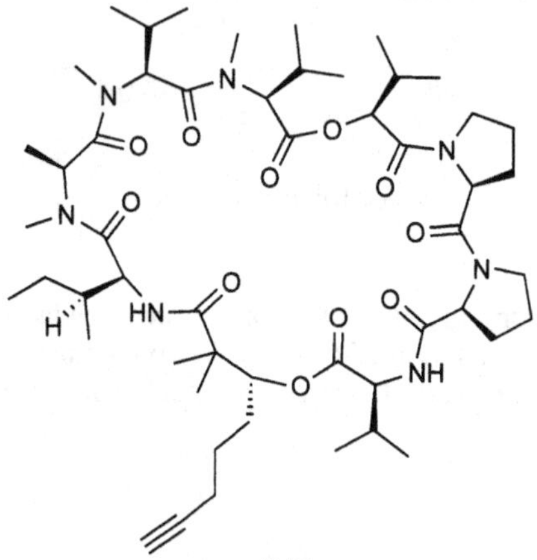

Figure 7.10 Cyclomarin A **7.31** and wewakpeptin A **7.32**.

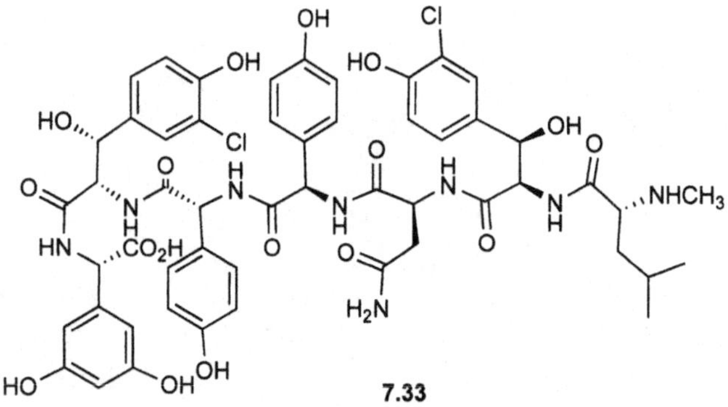

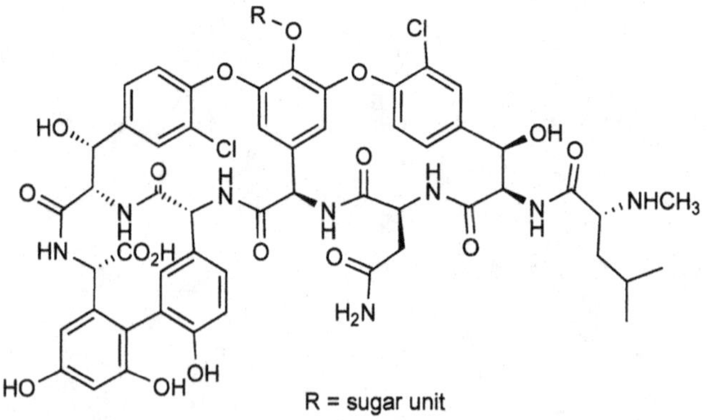

Figure 7.11 Vancomycin **7.34** and its precursor **7.33**.

7.4 Porphyrins

It is convenient to mention briefly the biosynthesis of porphyrins at this point because the amino acid, glycine, plays an important role in their biogenesis (Figure 7.12). Glycine condenses with pyridoxal phosphate **7.35** giving the imine **7.36** which reacts with the coenzyme A derivative **7.37** of succinic acid, giving the amino acid 5-aminolaevulinic acid **7.38**. Self-condensation of two molecules of compound **7.38** affords porphobilinogen **7.39** – the basic building block of the porphyrins.

Tetramerization of compound **7.39** giving uroporphyrinogen III **7.40** is catalyzed by two enzymes, deaminase and cosynthetase (Figure 7.13). Note the alternating arrangement of the acetic acid (–CH_2CO_2H) and propionic acid (–$CH_2CH_2CO_2H$) chains in rings A, B and C, and note how these chains in ring D are *turned around*. Compound **7.40** is then transformed in several steps into protoporphyrin IX **7.41**, which is the key intermediate in chlorophyll and haem biosynthesis.

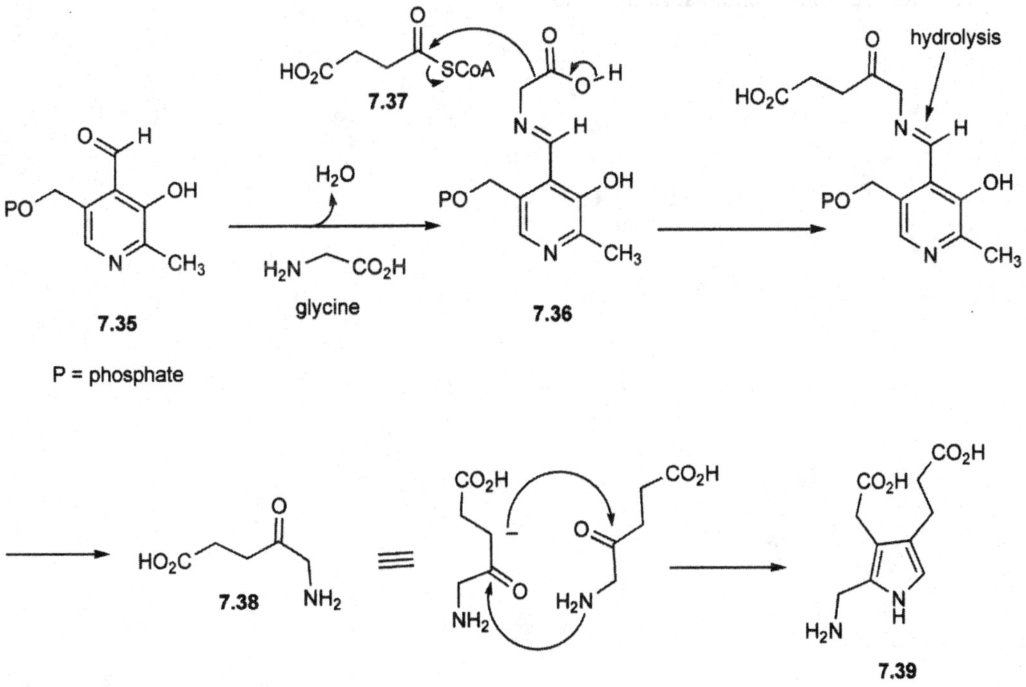

Figure 7.12 Biosynthesis of porphobilinogen **7.39**.

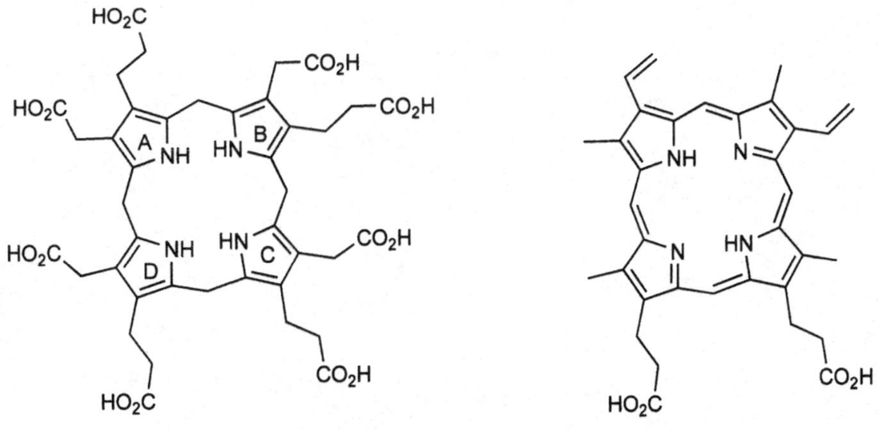

Figure 7.13 Uroporphyrinogen III **7.40** and protoporphyrin IX **7.41**.

Summary

(a) Amino acids are the biosynthetic precursors to many aliphatic alkaloids and macrocyclic molecules.
(b) Penicillins are derived from a tripeptide intermediate.
(c) Oxidative cyclization (penicillin) or cyclodehydration (thienomycin) reactions provide β-lactam rings.
(d) Porphyrins are derived from 5-aminolaevulinic acid.

Answers to Problems

Outline answers to the problems are given below. Note that in many multi-stage biosynthetic reactions the exact sequence of steps has not been determined, so you may well have a different sequence of reactions to those shown below. For the purpose of understanding biosynthetic routes to natural products, the knowledge of the exact order of reactions is not important.

Problem 3.1

See Figures 3.17 and 3.20 respectively.

Problem 3.2

There are two possible structures. They are structure **3.8** with (i) $R^1 = R^2 =$ stearic acid chain; $R^3 =$ oleic acid chain or (ii) $R^1 = R^3 =$ stearic acid chain; $R^2 =$ oleic acid chain.

Problem 3.3

Syn elimination.

Problem 3.4

Problem 3.5

Fatty acid **3.25** could be formed from an acetyl coenzyme A starter unit and then addition of 4 methylmalonyl coenzyme A units.

Problem 3.6

Problem 3.7

Compound **3.4**: Desaturation of stearic acid **3.1** giving oleic acid **3.2** followed by methylation as shown in Figure 3.11, pathway a.

Compound **3.5**: Desaturation of stearic acid **3.1** giving oleic acid **3.2** followed by cyclopropane ring formation as shown in Figure 3.11, pathway b. Further oxidative desaturation of the cyclopropane ring would give the cyclopropene ring of compound **3.5**.

Problem 3.8

Problem 4.1

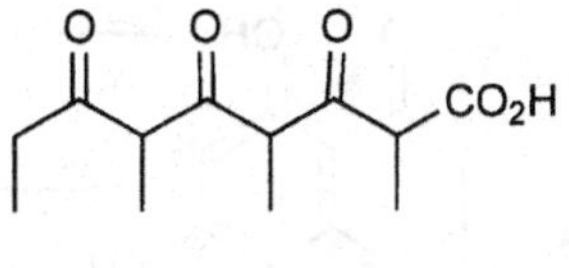

Problem 4.2

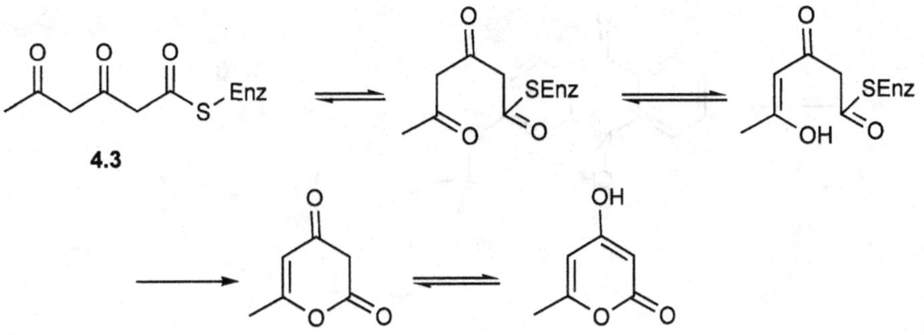

Problem 4.3

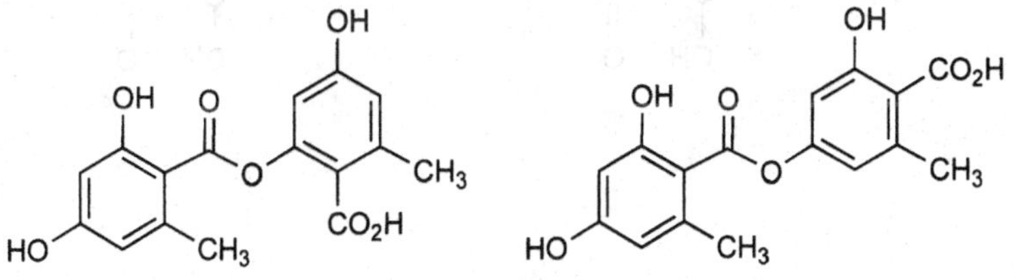

lecanoric acid

Problem 4.4

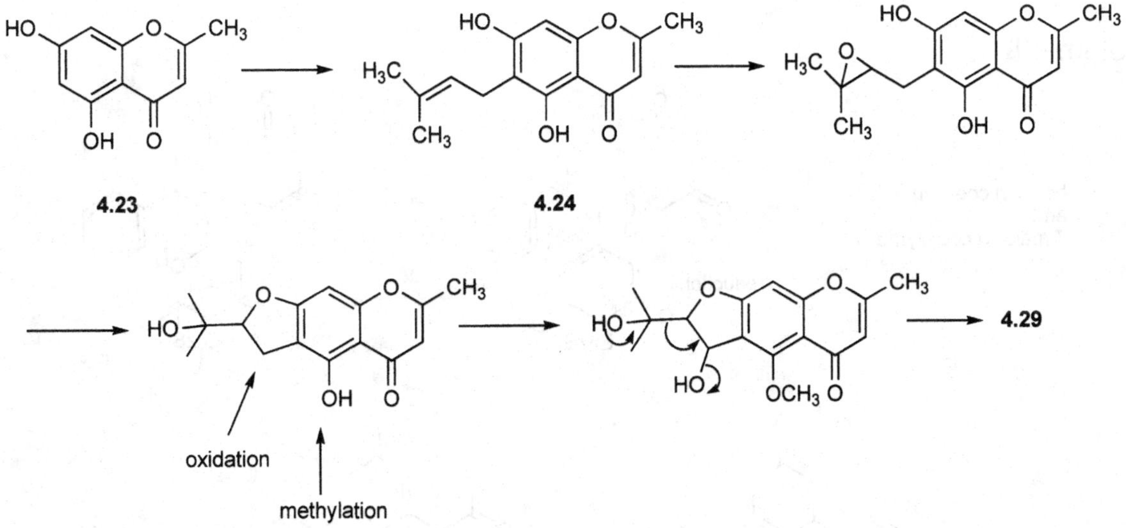

4.23　　　　　　　　　**4.24**

oxidation

methylation

4.29

Problem 4.5

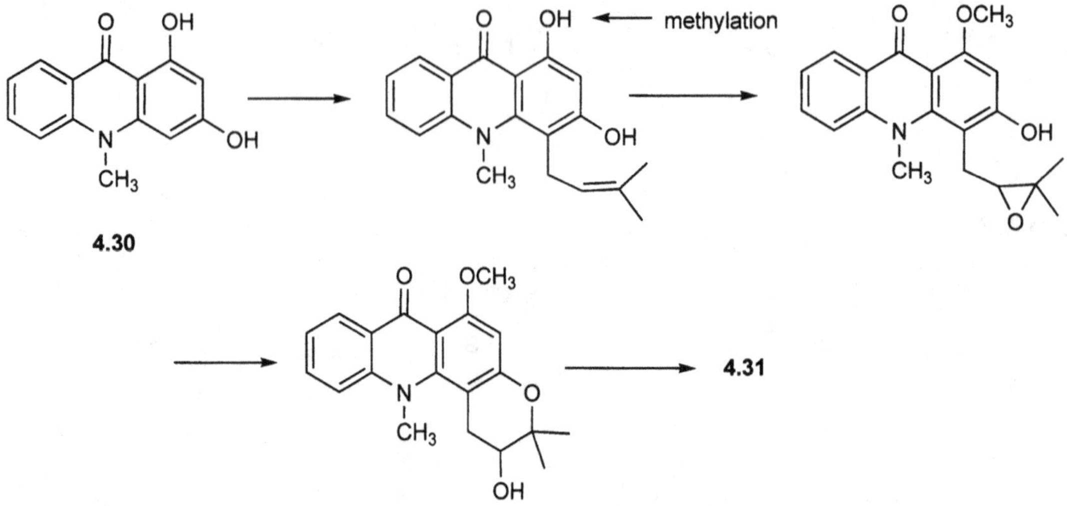

O
OH
OH
N
CH3
4.30
methylation
OCH3
4.31

Problem 4.6

SEnz
CH3
HO
OH
4.17
4.23

Problem 4.7

Enz
S
4.3
reduction then
dehydration
SEnz
malonyl
coenzyme A
4.40

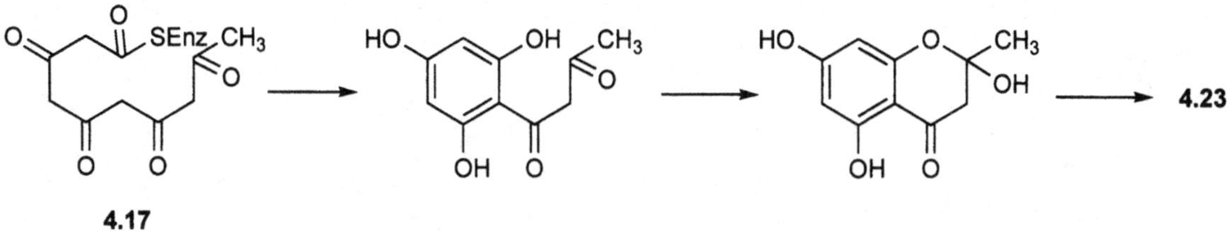

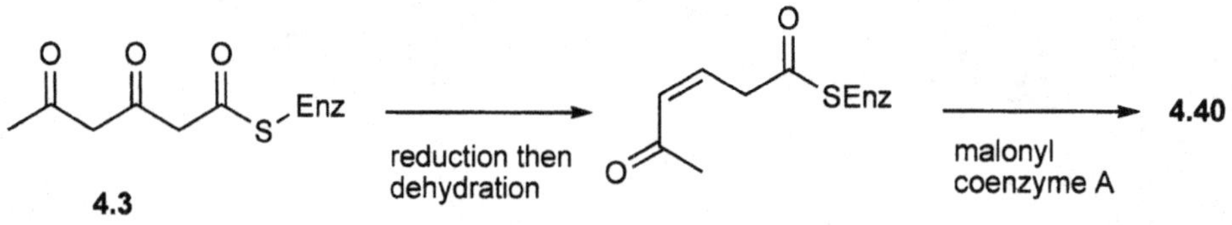

Problem 4.8

benzoyl coenzyme A
and
7 malonyl coenzyme A
reduction
EnzS
OH
4.42

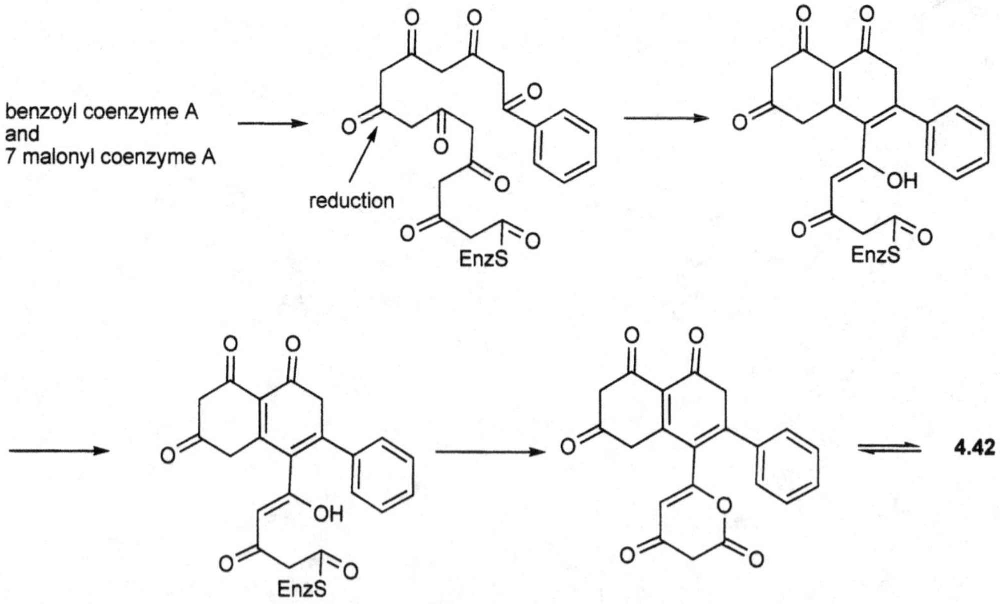

Problem 4.9

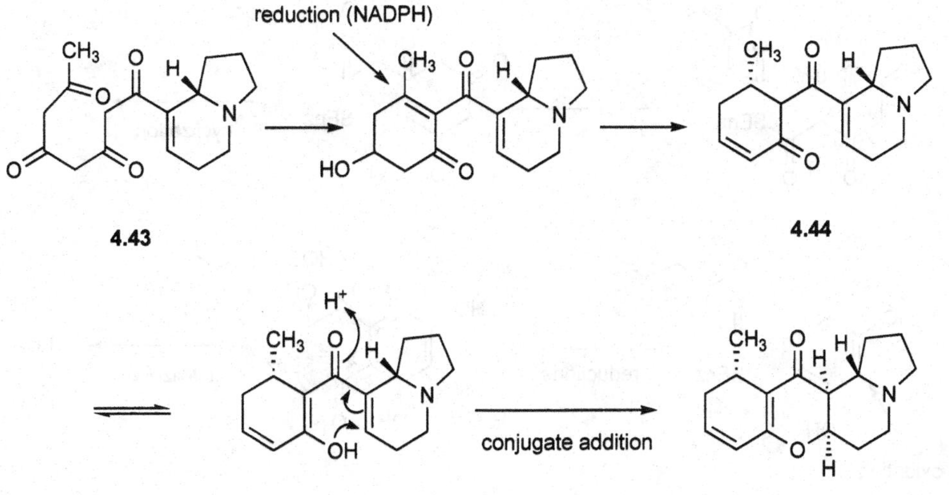

4.43

reduction (NADPH)

4.44

conjugate addition

4.45

Problem 4.10

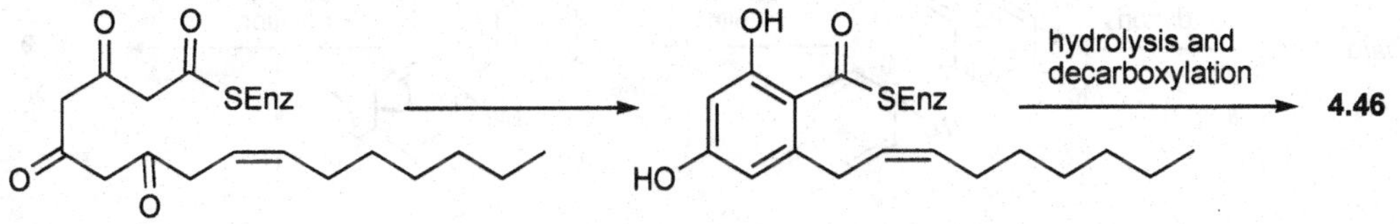

hydrolysis and decarboxylation **4.46**

Problem 4.11

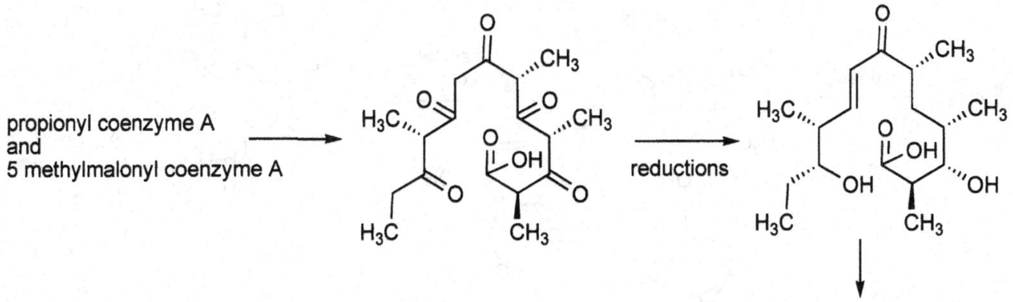

propionyl coenzyme A
and
5 methylmalonyl coenzyme A

reductions

4.53

Problem 4.12

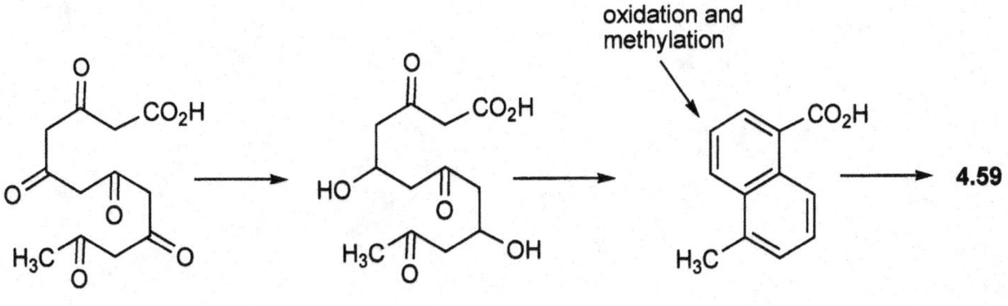

oxidation and methylation

4.59

Problem 4.13

Problem 4.14

Problem 4.15

Problem 4.16

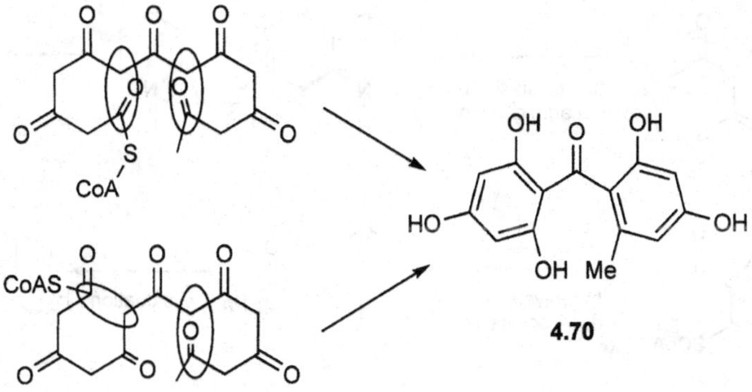

CoA
S
CoAS
OH
OH
HO
OH
OH
Me
4.70

Problem 4.17

oxidation
SEnz
3 x malonyl coenzyme A
HO
SCoA
4.73
oxidation
OH
OH
radical coupling
then O-methylations
4.74

Problem 4.18

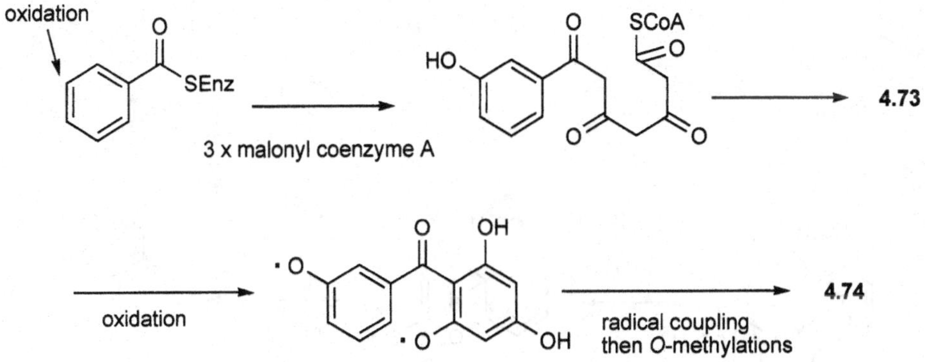

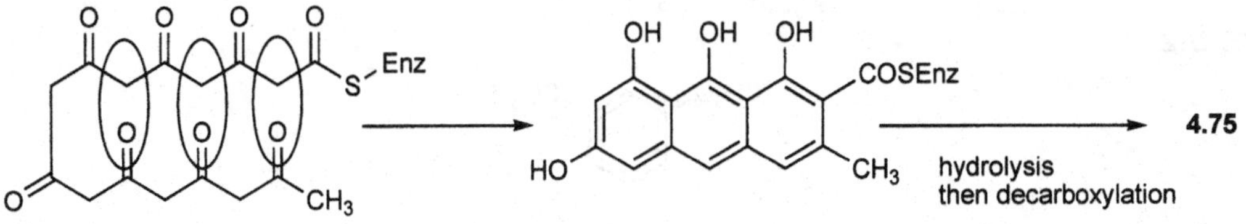

Enz
S
CH3
OH
OH
OH
COSEnz
HO
CH3
hydrolysis
then decarboxylation
4.75

Problem 4.19

OHC
H2NCH3
OHC
OH
HN
CH3
HO
OH
N
CH3
OH
N
CH3
HO2C
CO2H
4.86
decarboxylation

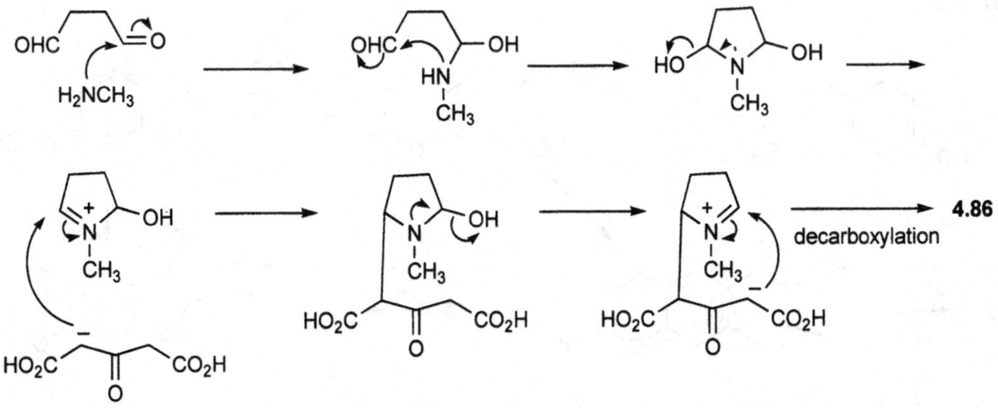

Problem 4.20

Problem 5.1

Problem 5.2

vanillic acid

Problem 5.3

Problem 5.4

Problem 5.5

Problem 5.6

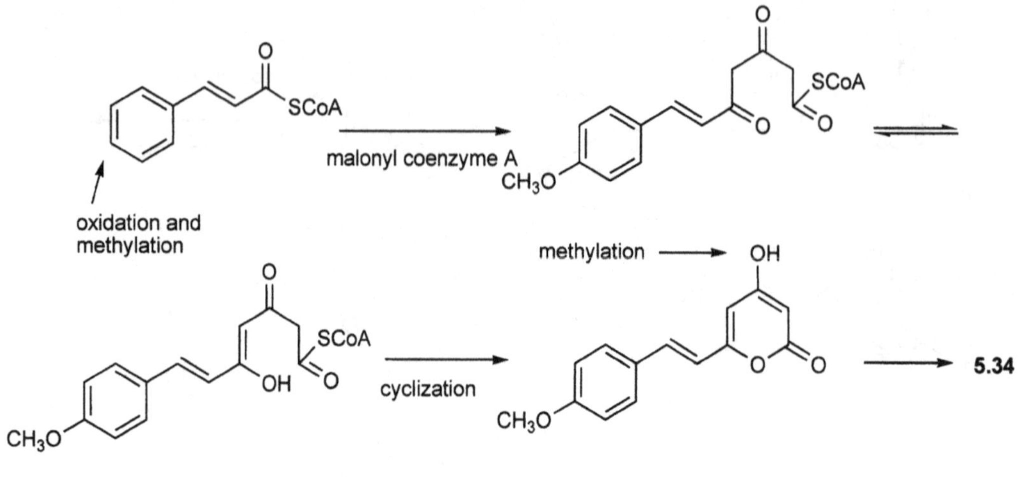

oxidation and
methylation

malonyl coenzyme A

methylation

cyclization

5.34

Problem 5.7

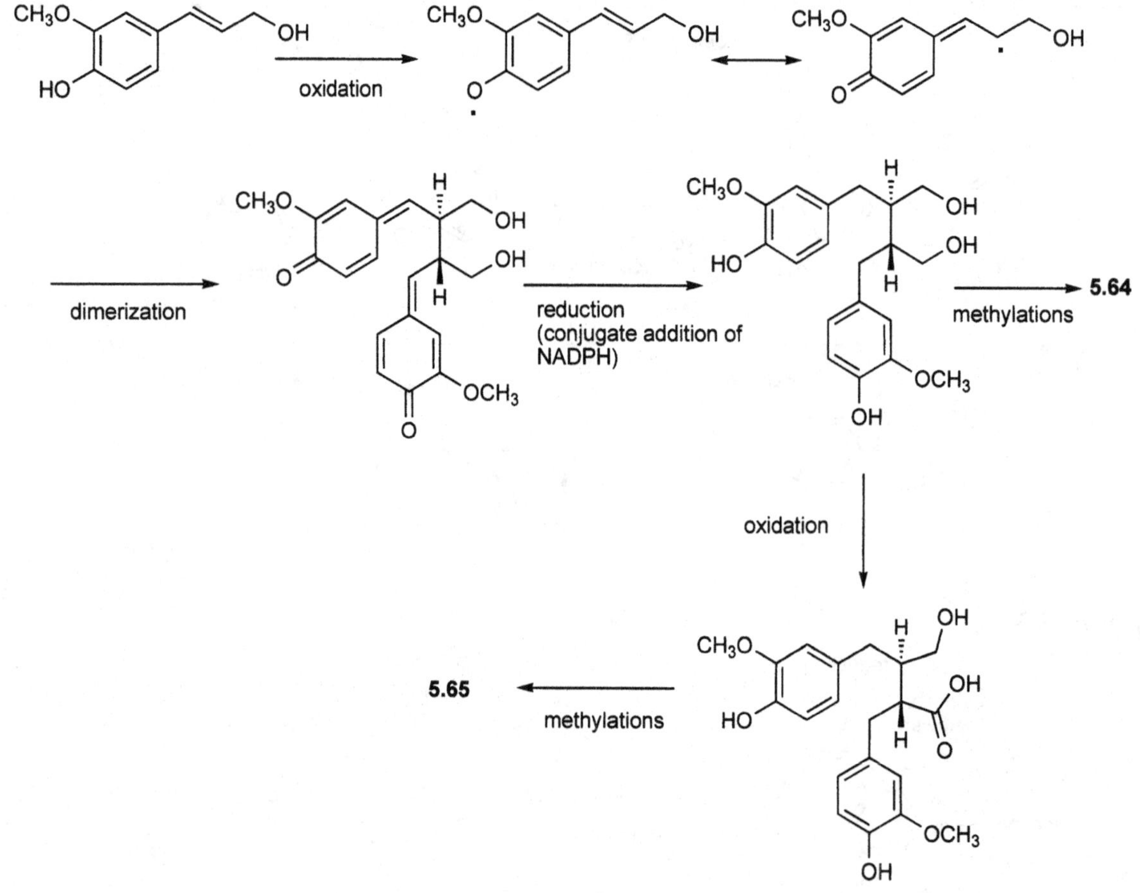

oxidation

dimerization

reduction
(conjugate addition of
NADPH)

methylations

5.64

oxidation

5.65

methylations

Problem 5.8

phosphorylation

reduction (NADPH)

isoeugenol

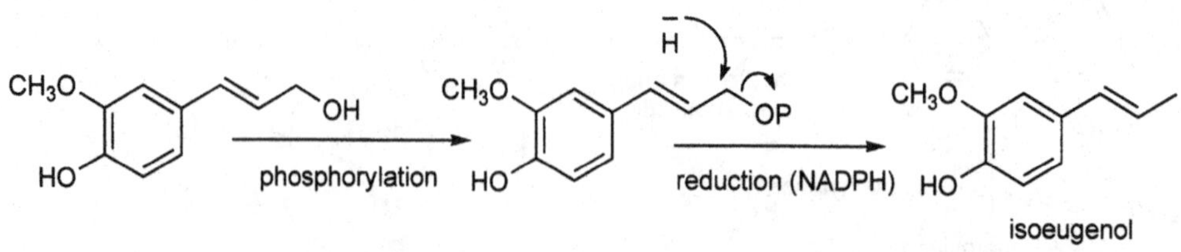

Problem 5.9

Problem 5.10

Problem 5.11

Problem 5.12

Problem 5.13

Problem 5.14

Problem 5.15

Problem 5.16

Problem 6.1

Problem 6.2

Problem 6.3

Problem 6.4

Problem 6.5

Problem 6.6

Problem 6.7

Problem 7.1

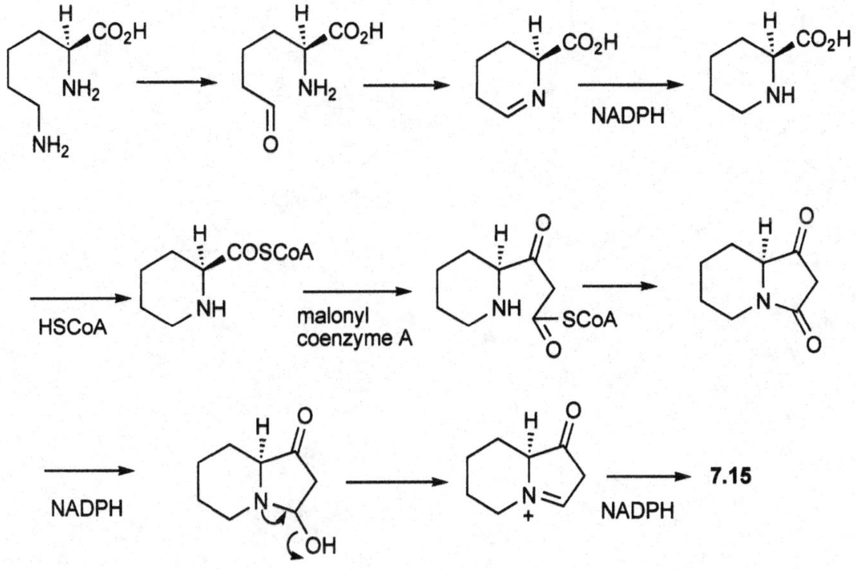

Problem 7.2

Tyrosine

Problem 7.3

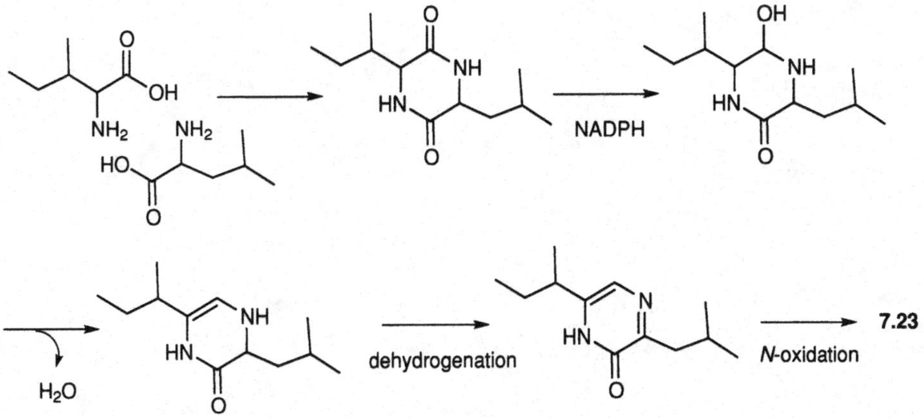

135

Further Reading

General Texts

Mann, J. (1994) *Chemical Aspects of Biosynthesis*; Oxford.
This book also contains a section on ecological chemistry.

Mann, J. (1987) *Secondary Metabolism* (2nd edition); Oxford.
This book also contains a section on ecological chemistry.

Dewick, P. M. (2002) *Medicinal Natural Products: a Biosynthetic Approach* (2nd edition); Wiley.
This book includes substantial discussions of the pharmacological and medicinal properties of natural products.

Torssell, K. B. G. (1983) *Natural Product Chemistry: a Mechanistic and Biosynthetic Approach to Secondary Metabolism*; Wiley.

Stanforth, S. P. *Natural Product Chemistry* (an open learning module); http://www.nl-webspace.co.uk/~unn_chss1/

Chemical Ecology

Crombie, L. (1999) Natural Product Chemistry and its part in the Defence Against Insects and Fungi in Agriculture. *Pesticide Science*, **55**, 761–774.

Wink, M. (2003) Evolution of Secondary Metabolites from an Ecological and Molecular Phytogenetic Perspective. *Phytochemistry*, **64**, 3–19.

Simmonds, M. S. J. (2003) Flavonoid-Insect Interactions: Recent Advances in our Knowledge. *Phytochemistry*, **64**, 21–30.

Fatty Acids

Hatanaka, A. (1993) The Biogeneration of Green Odour by Green Leaves. *Phytochemistry*, **34**, 1201–1218.

Blée, E. (1998) Phytooxylipins and Plant Defense Reactions. *Progress in Lipid Research*, **37**, 33–72.

O'Hagen, D., Harper, D. B. (1999) Fluorine-containing Natural Products. *Journal of Fluorine Chemistry*, **100**, 127–133.

Dembitsky, V. M., Srebnik, M. (2002) Natural Halogenated Fatty Acids: their Analogues and Derivatives. *Progress in Lipid Research*, **41**, 315–367.

Rezzonico, E., Moire, L., Poirier, Y. (2002) Polymers of 3-Hydroxyacids in Plants. *Phytochemistry Reviews*, **1**, 87–92.

Behrouzian, B., Buist, P. H. (2003) Bioorganic Chemistry of Plant Lipid Desaturation. *Phytochemistry Reviews*, **2**, 103–111.

Kunst, L., Samuels, A. L. (2003) Biosynthesis and Secretion of Plant Cuticular Wax. *Progress in Lipid Research*, **42**, 51–80.

Polyketides

Fujii, I., Ebizuka, Y. (1997) Anthracycline Biosynthesis in *Steptomyces galilaeus*. *Chemical Reviews*, **97**, 2511–2523.

Staunton, J., Wilkinson, B. (1997) Biosynthesis of Erythromycin and Rapamycin. *Chemical Reviews*, **97**, 2611–2629.

Shikimic Acid Pathway

Umezawa, T. (2003) Diversity in Lignan Biosynthesis. *Phytochemistry Reviews*, **2**, 371–390.

Terpenes

Rohmer, M. (2003) Mevalonate-independent Methyerythritol Phosphate Pathway for Isoprenoid Biosynthesis. Elucidation and Distribution. *Pure and Applied Chemistry*, **75**, 375–387.

Wise, M. L. (2003) Monoterpene Biosynthesis in Marine Algae. *Phycologia*, **42**, 370–377.

General

Hadacek, F. (2002) Secondary Metabolites as Plant Traits: Current Assessment and Future Perspectives. *Critical Reviews in Plant Sciences*, **21**, 273–322.

Index